George Englebretsen
Figuring It Out

Philosophical Analysis

Edited by
Rafael Hüntelmann, Christian Kanzian, Uwe Meixner,
Richard Schantz, Erwin Tegtmeier

Volume 78

George Englebretsen
Figuring It Out

Logic Diagrams

In Cooperation with José Martin Castro-Manzano and
José Roberto Pacheco-Montes

DE GRUYTER

ISBN 978-3-11-076335-5
e-ISBN (PDF) 978-3-11-062445-8
e-ISBN (EPUB) 978-3-11-062184-6
ISSN 2627-227X

Library of Congress Control Number: 2019951086

Bibliographic Information published by the Deutsche Nationalbibliothek
The Deutsche Nationalbibliothek lists this publication in the Deutsche Nationalbibliografie; detailed bibliographic data are available on the Internet at http://dnb.dnb.de.

© 2021 Walter de Gruyter GmbH, Berlin/Boston
This volume is text- and page-identical with the hardback published in 2020.
Druck und Bindung: CPI books GmbH, Leck

www.degruyter.com

For:
Libbey, who draws the lines
Russell, who keeps the lines straight
Suzanne, who minds the intersections
Morgan, who directs the arrows
and Gaël, who still draws the conclusion
(Emma, Carlee, Sam, Nathan, and Aaron continue to run in circles)

Contents

List of Figures —— IX

Preface —— XIII

1 Introduction: Seeing Reason —— 1

2 Some Historical Figures —— 8
2.1 Aristotle and the Lost Diagrams —— 8
2.2 "The Golden Age" – Leibniz, Lambert, and Euler —— 26
2.3 Venn, Peirce (and Frege?) —— 34
2.4 "The Renaissance of Diagrammatology" —— 54

3 Lines of Reason —— 58
3.1 New Lines (Smyth and Pagnan) —— 58
3.2 In Logical Terms: Term Functor Logic —— 66
3.3 Term Lines —— 75
3.4 The Point of Names —— 98
3.5 Vectors of Relations —— 105
3.6 World Lines —— 123

4 Holding the Line —— 139
4.1 Figuring it Out —— 139
4.2 Crossing the Line —— 145
4.3 Seeing Reason —— 152
4.4 So, How Do We Reason, After All? —— 157

5 Linear Diagrams and Non-Classical Quantifiers —— 163
 José Martin Castro-Manzano and José Roberto Pacheco-Montes
5.1 Introduction —— 163
5.2 Two Frameworks for Syllogistic —— 163
5.2.1 General aspects of syllogistic —— 163
5.2.2 The TFL framework: the plus-minus algebra —— 164
5.2.3 The SYLL$^+$ framework: the extra quantifiers —— 166
5.3 The TFL$^+$ Framework: A Tweaked Version of Syllogistic —— 170
5.3.1 Step 1. The plus-minus algebra meets the extra quantifiers —— 171
5.3.2 Step 2. The plus-minus algebra modification —— 172
5.3.3 Step 3. Reliability —— 174

5.4	The TFL$^⊕$ Framework: A Diagrammatic Extension —— **176**	
5.4.1	Step 1. The linear diagrams meet the extra quantifiers —— **178**	
5.4.2	Step 2. The procedure modification —— **180**	
5.4.3	Step 3. Reliability —— **182**	
5.5	Conclusions —— **185**	
	Appendix: Rules of inference for TFL —— **186**	
6	**Drawing Conclusions: Get Better Diagrams —— 188**	

Bibliography and Further Reading —— 194

Index —— 214

List of Figures

Figure 2.1: The Pythagorean Theorem —— 9
Figure 2.2: The Pentacle and the Golden Ratio —— 10
Figure 2.3: π —— 11
Figure 2.4: The Diagonal of a Unit Square —— 11
Figure 2.5: Aristotle's First Square —— 15
Figure 2.6: Aristotle's Second Square —— 15
Figure 2.7: Possible Line Diagrams for Barbara and Celarent —— 17
Figure 2.8: Diagram of AB —— 23
Figure 2.9: Diagrams for the Three Aristotelian Figures —— 23
Figure 2.10: Diagram for Cesare —— 23
Figure 2.11: Barbara and Baroco —— 24
Figure 2.12: Leibniz's Linear Diagrams of the Categoricals (extensional) —— 28
Figure 2.13: Diagram of Barbara —— 29
Figure 2.14: Leibniz's Linear Diagrams of the Categoricals (intensional) —— 30
Figure 2.15: Lambert Diagrams —— 31
Figure 2.16: Euler Diagrams of the Categoricals —— 33
Figure 2.17: Euler's Asterisk for 'Some S is P' —— 33
Figure 2.18: Euler Diagram for Cesare —— 34
Figure 2.19: Euler Diagrams for 'Some A is not B' and 'Some B is not A' —— 35
Figure 2.20: Euler Diagrams for Contradictory Pairs —— 36
Figure 2.21: Euler Diagrams for 'All A are B' and 'Some A are C' —— 36
Figure 2.22: Euler Diagrams for Contrary Pairs —— 37
Figure 2.23: Venn Diagrams of Celarent and Festino —— 38
Figure 2.24: Square Diagram for 4 Terms —— 40
Figure 2.25: Peirce's Diagram for Festino —— 41
Figure 2.26: Pierce Diagram of Propositional Disjunction —— 41
Figure 2.27: Alpha Graphs —— 43
Figure 2.28: Beta Graphs for the Four Categoricals —— 44
Figure 2.29: Beta Graphs with Relationals —— 45
Figure 2.30: Beta Graphs of Identity —— 45
Figure 2.31: Beta Graphs for Quantification —— 46
Figure 2.32: Beta Graphs Illustrating Quantifier Scope —— 46
Figure 2.33: A Single Beta Graph Illustrating a Logical Equivalence —— 47
Figure 2.34: Fregean Diagrams for Propositional Logic —— 49
Figure 2.35: Fregean Diagrams for Predicate Logic —— 50

Figure 2.36: Fregean Diagram —— 50
Figure 2.37: Fregean Modus Ponens —— 51
Figure 2.38: Fregean Universal Instantiation —— 52
Figure 2.39: A Fregean Proof —— 53
Figure 3.1: Directed Graphs of A, E, I, and O —— 59
Figure 3.2: Directed Graph of an Argument —— 59
Figure 3.3: Directed Graph of Barbara —— 60
Figure 3.4: Directed Graph of Celarent —— 60
Figure 3.5: Directed Graph of Darii —— 61
Figure 3.6: Directed Graph of Ferio —— 61
Figure 3.7: SYLL Diagrams of the Four Perfect Syllogisms —— 63
Figure 3.8: SYLL Diagram of Baramtip —— 64
Figure 3.9: SYLL Proof that E and I are Contradictory —— 64
Figure 3.10: A Term Line —— 78
Figure 3.11: A Singular Term Point —— 78
Figure 3.12: A Negative Term Line —— 78
Figure 3.13: Universal Affirmative Line Diagram —— 79
Figure 3.14: Line Diagram for Two Co-extensive Terms —— 79
Figure 3.15: Universal Negative Line Diagram —— 80
Figure 3.16: Particular Affirmative Line Diagram —— 80
Figure 3.17: Diagrams for Predicate Denial and Term Negation —— 80
Figure 3.18: Line Diagrams for Obverted Categoricals —— 81
Figure 3.19: Line Diagrams for Affirmative and Negative Singulars —— 81
Figure 3.20: An Attempted Diagram for a Contradictory Pair —— 83
Figure 3.21: Diagrams for Tautologies —— 83
Figure 3.22: Full Diagram of A —— 84
Figure 3.23: Full Diagram of E —— 84
Figure 3.24: Full Diagram of I —— 85
Figure 3.25: Full Diagram of O —— 85
Figure 3.26: Contrapictions of A/E Contrariety —— 86
Figure 3.27: Diagram of Subcontrariety —— 86
Figure 3.28: Two Versions of Subalternation —— 88
Figure 3.29: Line Diagrams for the Perfect Syllogisms —— 93
Figure 3.30: Line Diagrams for Universal and Particular Syllogisms —— 93
Figure 3.31: 'Every M is P' —— 93
Figure 3.32: Barbara —— 94
Figure 3.33: Cesare 1 —— 94
Figure 3.34: Cesare 2 —— 94
Figure 3.35: Datisi —— 94
Figure 3.36: Bocardo —— 95

Figure 3.37: Ferison —— 95
Figure 3.38: Fresison —— 95
Figure 3.39: Barbarip —— 96
Figure 3.40: A Five-term Diagram —— 97
Figure 3.41: An Inconsistent Set —— 97
Figure 3.42: A Contrapiction of Baroco —— 98
Figure 3.43: Wild Quantity —— 99
Figure 3.44: Diagrams for Singular and Negative Singular Subjects —— 101
Figure 3.45: Leibniz's Syllogism with a Singular Predicate Term —— 102
Figure 3.46: Diagrammed Syllogism with Two Singular Terms —— 102
Figure 3.47: Reflexivity, Symmetry, and Transitivity in TFL —— 103
Figure 3.48: Diagrams With Terms of Explicit Denotation —— 105
Figure 3.49: Diagrams With a Team Name and With an Explicit Team Name —— 105
Figure 3.50: Romeo loves Juliet —— 106
Figure 3.51: Juliet loves Romeo —— 107
Figure 3.52: Unanalyzed Complex Relational —— 108
Figure 3.53: Partially Analyzed Complex Relational —— 108
Figure 3.54: Fully Analyzed Complex Relational —— 109
Figure 3.55: The Principle of Relational Analysis —— 109
Figure 3.56: A Corollary —— 110
Figure 3.57: Unsimplified Relational —— 110
Figure 3.58: Simplified Relational —— 110
Figure 3.59: Valid Relational Inference —— 111
Figure 3.60: De Morgan's Inference —— 111
Figure 3.61: Four Special Inferences —— 112
Figure 3.62: The Principle of Relational Extension —— 113
Figure 3.63: Specific vs Non-Specific Reference —— 114
Figure 3.64: The Principle of Relational Reduction —— 115
Figure 3.65: Attempted Singular Reflexive Diagram —— 115
Figure 3.66: Singular Reflexive Diagram —— 116
Figure 3.67: Attempted General Diagram —— 116
Figure 3.68: Proper Diagram —— 117
Figure 3.69: Argument With a Reflexive —— 117
Figure 3.70: A Simple Pronominalization —— 119
Figure 3.71: A Full Pronominalization —— 119
Figure 3.72: An Example of Interlocking Pronominalizations —— 120
Figure 3.73: A Diagrammed Inference with Pronouns —— 121
Figure 3.74: Negative Relationals —— 122
Figure 3.75: Inference with Negative and Positive Relationals —— 123

Figure 3.76: How to Diagram a Singular —— **129**
Figure 3.77: Diagrams for Sentential Logic —— **131**
Figure 3.78: The Domain of Domains for Conditionals —— **131**
Figure 3.79: Alternative Diagrams for a Conjunctive Statement —— **131**
Figure 3.80: Diagrams for Contraposition —— **132**
Figure 3.81: Some Elementary Rules of Propositional Logic —— **133**
Figure 3.82: Unanalyzed and Analyzed Complex Terms —— **133**
Figure 3.83: A Diagrammed Deduction —— **134**
Figure 3.84: Diagram of a Consistent Set of Statements —— **135**
Figure 3.85: Proof of an Inconsistent Set of Statements —— **136**
Figure 3.86: Diagrammatic Resolutions of the Two Disanalogies —— **138**
Figure 4.1: Lemon and Pratt Counter-example —— **146**
Figure 4.2: Illustration of the Principle of Compound Term Analysis —— **147**
Figure 4.3: First Response to Lemon and Pratt —— **148**
Figure 4.4: Line Diagrams for 'a is a' —— **149**
Figure 4.5: Beta Graphs for 'Something is a' —— **149**
Figure 4.6: Second Response to Lemon and Pratt —— **149**
Figure 4.7: Beta Graph for 'A is greater than something greater than B' —— **151**
Figure 4.8: ED Diagram for 'A is greater than something greater than B' —— **151**
Figure 4.9: Beta Graph for 'A man gave a bribe to a senator' —— **151**
Figure 4.10: ED Diagram for 'A man gave a bribe to a senator' —— **151**
Figure 4.11: Beta Graph for Ferio —— **152**
Figure 4.12: ED Diagram for Ferio —— **152**
Figure 4.13: The Number Line —— **154**
Figure 5.1: Extended Square of Opposition Adapted From (Thompson 1986, p. 77) —— **168**
Figure 5.2: Syntax for Englebretsen's Linear Diagrams Adapted From (Englebretsen 1992a) —— **177**
Figure 5.3: Examples of Syllogisms With Linear Diagrams Adapted From (Englebretsen 1992a) —— **177**
Figure 5.4: Vocabulary for TFL$^\oplus$ —— **178**
Figure 5.5: Syntax for TFL$^\oplus$ —— **179**
Figure 5.6: A Valid Example: ekg-1 Diagram —— **181**
Figure 5.7: An Invalid Example: akt-4 Diagram —— **181**
Figure 5.8: A Valid Reasoning With Relations —— **182**
Figure 5.9: Valid Syllogistic Forms in TFL$^\oplus$ According to the TFL$^+$ Conditions —— **184**

Preface

You ought to start with Logic.

                                                                                                   Goethe

Possemus cum bacylis syllogizare et cum lapidibus concludere.

                                                                   Paul of Venice

Did you ever see such a thing as a drawing of a muchness?

                                                                     Lewis Carroll

I'm not bad. I'm just drawn that way.

                                                                    Jessica Rabbit

Let us draw instead of talk.

                                                                          Goethe

Bertrand Russell famously said that a good notation is like a live teacher. A good system of logical symbolization can contain valuable, but often hidden, riches of which even its author was unaware. Fred Sommers' plus/minus symbolic algorithm for an expanded, strengthened term logic (Term Functor Logic) turns out to be just that kind of "good notation". It is a system of formal logic that was the focus of much of my own research beginning half a century ago. Eventually, in the 1980s, my interest turned to the possibility of devising a system of logic diagrams that would be a graphic analogue to term logic in much the way that Venn meant his system of diagrams to be an analogue for Boole's logic. This led to many discussions with Sommers. He said that he had tried to do this for his new version of term logic but kept coming back to simple Venn, and then Peirce, diagrams. I thought something like Leibniz's linear diagrams might be a better place to start. After a number of failed attempts, it turned out that once I thought more deeply about Sommers' notation, keeping it as a kind of guide, I began to make progress. In 1992 I published my own system of linear diagrams, one that could serve as a viable medium for conducting and analyzing formal reasoning. This system could be used not only for a logic of monadic terms but of relational terms, singular terms, and even a logic of propositions as well. That work has received considerable attention during the last twenty-five years. Critiques and suggestions concerning it have accumulated to the extent that I began to work on a project to address these. I also felt that it would be appropriate to do a better job of situating the system of linear diagrams within the long, broad history of attempts at diagrammatic treatments of logic, and also to make clear Sommers' background logic of terms that inspired it. The present work is the result of these efforts as well as, of course, the impressive work of two important contributors to the field of logic diagrams, José Martin Castro-Manzano and José Roberto Pacheco-Montes, who have provided a fine example of that to be seen below.

Like any other researcher, I have relied upon and been helped by many others. But a few of them deserve special recognition. First, of course, to my dear long-time friend Fred Sommers, who died as I was writing this book. Fred was a philosopher-logician who pushed me along his path toward a revitalized, strengthened, natural, viable version of term logic that went far beyond the traditional term logic initiated by Aristotle, and that met the standards of expressive and inferential power demanded of any system of formal logic. As well, forty years ago he encouraged me to continue my then nascent attempts to formulate a linear system of diagrams for logic that would allow representations of relations. I also want to express my profound gratitude to Amirouch Moktefi, whose own research, insights, and encouragement have been invaluable over many years. I am happy for this opportunity to thank Marian Wesoły, a first-rate scholar of Aristotle's logic and who was the first to provide a reasonable and textually accurate reconstruction of Aristotle's lost syllogistic diagrams. He confirmed my own conviction that such diagrams were worth seeking and encouraged me in my own attempts at devising diagrams for syllogisms. John Corcoran's careful historical studies, especially of Aristotle's syllogistic, have been extremely useful to me for many years. Charles Hornbeck has been a generous and valuable sounding board for a number of my ideas, including those that eventually led to my work here. Several other scholars have offered me useful advice, helpful criticisms, or encouragement with this project. Among them are Hanoch Ben-Yami, Lorenz Demy, Danny Frederick, Brian Gaines, J.A. Laronge, Koji Mineshima, Mehdi Mirzapour, Christopher Morrissey, Larry Moss, Wallace Murphree, David Oderberg, Ahti-Veikko Pietarinen, Yuri Sato, Fabien Schang, Topal Selçuk, Sun-Joo Shin, and the late Lorne Szabolcsi, Philip Peterson, and B. Hartley Slater. I thank them all and blame them for nothing. Holly McMillan has yet again shown that she is the Queen (and her assistant Lizzie, the Princess) of manuscript preparation. Lastly, but sincerely and with ever-lasting love, I thank Libbey, Russell, Suzanne, Morgan, and Gaël, to whom I have dedicated this book.

1 Introduction: Seeing Reason

Logical Diagram (or Graph): Ger. *logische Figur*; Fr. *diagramme logique*; Ital. *diagramma logico*. A diagram composed of dots, lines, &c., in which logical relations are signified by such spatial relations that the necessary consequences of these logical relations are at the same time signified, or can, at least, be made evident by transforming the diagram in certain ways which conventional 'rules' permit.

J.M. Baldwin (1901)

Arithmetical symbols are written diagrams and geometrical figures are graphic formulas.

D. Hilbert

All notation, no doubt, is both pictorial and arbitrary: nevertheless there are cases in which one or the other character decidedly predominates ... the syllogism admits of a graphical representation which is as suggestive as a diagram in geometry.

De Morgan

Par le moïen de ces signes tout saute d'abord aux yeux.

Euler

I have always tended to favour the visual over the auditory. Perhaps because blindness has coursed through my family. The visible is often worth cherishing. But there is at least another reason: sometimes, seeing is (the most direct path to) believing.

Around 3500bce, a few people in Mesopotamia invented cuneiform writing by scratching marks on pieces of clay. By 3100bce, a few people in Egypt developed hieroglyphic writing by carving marks on stone. Writing, making visible what is said, was a skill confined, at least in those times, to a very few, very special, people. It was a skill difficult to acquire. Those few who had such a skill were often viewed with awe or fear. Writing was some kind of magic. Its products were beyond the understanding of most people. Writings held secrets. Today, the ability to read and write, literacy, is a bit more common. One rarely thinks consciously about it – even when exercising it. We've learned to read (and, soon after, to write) our native language. In doing so, we adopted a number of conventions generally accepted by the members of our linguistic community. We tacitly agree that this series of marks must be produced only in certain specifiable written contexts, that this series of marks must not be augmented by this or that specified mark, etc. (rules of spelling, punctuation, grammar, and so forth). All these are matters of convention.

One of the many, but perhaps the most important, reasons we speak is to convey information to others. Writing gave us two good things: a second means of doing most of the things we do with speaking, and a means of preserving what was or might have been said. The written word may be cheap and fleeting (especially in our digital age), but the spoken word is even cheaper and more fleeting. Writing was a very good idea. Suppose I wish to convey to you the information that not all the male board members at yesterday's board of directors' meeting were wearing ties. I could simply tell you by saying something like, 'Some guy wore no tie' (this must be some kind of high-end, very formal Wall Street firm). Presumably, you understand, without it being made explicit, that I'm talking about just the men who attended the pertinent meeting. But you are not readily available now. So I write to you. I write something like, 'Some guy wore no tie' or 'There was a director there without a tie' or even simply 'They didn't all have ties'. It doesn't much matter which writing medium I use. I could email you or even write it on stationery, put it in an envelope, stamp it, and mail it to you (if that's anything people younger than me still do). However I do it, if you successfully receive and understand my missive, then I have conveyed to you the information I wanted to convey: that at least one male board member at yesterday's meeting of the board of directors did not wear a tie. Given that I can take for granted with a high degree of certainty that you know who all the directors are, and which meeting of them I have in mind, I might even dispense with writing anything at all. I might very well simply send you a photograph of the people sitting around that big expensive table in the boardroom. Do you see that guy? What's he thinking? If I want to make sure you see which one I mean, I might draw an arrow pointing to him, or I might just write his name over his image, or I might draw a smiley face above it. I've used the picture to convey to you just the information I wished to convey. But I didn't write anything (almost). This picture really could be worth a thousand (or at least a half-dozen or so) words.

Seeing Bill and Hillary, we *see that* Bill is taller than Hillary. We could represent this information, assuming we care only about their relative heights, by drawing stick figures of different lengths and labeling one as 'B' (for Bill) and the other as 'H' (for Hillary). The drawing is a kind of *diagram*. Anyone who knows the labelling convention and the line length convention, can view the diagram and conclude (i.e., reason on the basis of it alone) that Bill is taller than Hillary. Of course, we need not use any such diagram. We could (and probably would) simply say, or write, that Bill is taller than Hillary by using an appropriate sentence in English, or any other natural language (e.g., French or Farsi). We could express this information even in a non-natural, artificial language, one formulated for the purposes of reflecting and facilitating how we draw conclusion from previously acquired information. We might make use of our old labels, 'B' and 'H', and then introduce a symbol for the relation of *being taller than*. We could then adopt a convention determining how, and in which order, we write these labels and symbols. Anyone familiar with such a symbolic language could then just read the string of symbols and thereby understand that Bill is taller than Hillary.

So what? It's all pretty obvious and trivial. What's Logic got to do with it? Well, let's start with mathematical formulae. Mathematicians make use of a special non-natural, formal, artificial language to preserve and convey mathematical information. Lots of people can use this language, not only mathematicians but all kinds of scientists and other STEM types, not to mention accountants and others, even a few of those board members at the bank. Still most people have at best a passing familiarity with the language of mathematics. For many, basic arithmetic and $E=mc^2$ are enough. People who can read and write the language of mathematics are, like their ancient Mesopotamian and Egyptian ancestors were, viewed with awe and (sometimes) fear. They practice some kind of magic. What they write is beyond the understanding of most people. Mathematical language holds secrets. Well maybe that's not true of all those mathaphobes. Geometry at least has pictures. And Logic?

Logic is not Mathematics. It just wears mathematical clothes now and then. For most of the time Logic has been around, at least in the West, it has been carried out in one or more natural languages (first Greek, then Latin, then a variety of European languages, now mostly English). In the 19th century, a number of logicians thought of logic as a branch of mathematics (namely as a kind of algebra). By the end of the century logicians began thinking of logic as a good foundation for mathematics (which had been having troubles keeping its old foundations solid). They no longer made use of any natural language as the language of logic. Logicians began inventing their special non-natural, formal, artificial language. This time it was not the language of mathematics. Even after the idea that mathematics could rest on logic came to look fairly hopeless, logicians continued to use (and then refine and extend) their new language. Logic isn't taken any more to be mathematics, not even just a branch of mathematics or the foundation of mathematics (but it does proudly (?), defiantly (?) call itself 'Mathematical Logic' or 'Modern Mathematical Logic' or, more mysteriously 'First-Order Predicate Calculus', or more vaguely 'Symbolic Logic'). Still, it does look like it wears mathematical clothes. That's because it uses special symbols. Time for an illustration. The logician demands that before natural language sentences are fit to play the various roles that logicians are interested in (being premises or conclusions of inferences, being steps in a deduction, being members of a statement-set in the dock for inconsistent exposure, and so forth) it must be translated into the new symbolic language. Thus the sentence 'Some board member did not wear a tie' might be translated ("formulated") as: $(\exists x)(Bx . \sim Tx)$, Were you to ask the logician what this says in English, the response would probably be something like: 'There exists (in the universe of discourse) at least one thing such that both it is a Board member and it is not a Tie-wearer'. Of course a number of short-hands are available here. Like any natural language, any non-natural language relies for its use on a relatively large variety of conventions. By contrast, things like pictures, maps, charts, diagrams, etc. rely (in various degrees) on fewer (perhaps none, as in the case of my unmarked photo of the board members) conventions. While all logicians have focused on sentences (natural or otherwise) as fit to play the logical roles mentioned above, some logicians have also turned some of their attention to non-sentential, picture-like, information conveying devices – logic diagrams. An early example of this is called Euler diagrams. The Euler diagram for our sentence about a tieless board member would be a pair of overlapping circles with a 'B' inscribed in the part of one circle outside the overlapping section and a 'T' inscribed above the other circle. That's it. Just one look tells you that at least one thing that is B (a board member), indicated by the inscribed 'B', is not in the circle representing tie-wearers, a circle indicated by the inscription 'T'. Note that while my original sentence relied on a number of conventions (of spelling, grammar, etc.) and my photo relied on virtually no salient convention, the Euler diagram requires a small number of agreed upon conventions (for example, the 'B' stands for board members and 'T' stands for 'Tie-wearer', as well as the convention that the circles represent all the things that are board members in one case and tie-wearers in the other, and that the overlapping part of the circles represents things that are both board members and tie-wearers. I've just given a description. Yet surely at this point, for the novice at least, a picture, for example, the Euler diagram, might be appreciated. We will get to that, and many other logical diagrams, very soon. In the meantime, rest with the realization that you probably already have the Euler diagram I described. Using my description, you constructed a mental image of it – you *imagined* it.

Plato, as one would expect, had a higher regard for the figures we imagine than the ones we actually see. Thinking of geometers in particular, he wrote in *The Republic:*

> And do you not know also that although they make use of the visible forms and reason about them, that they are thinking not of these but of the ideal which they resemble; not of the figures which they draw, but of the absolute square and the absolute diameter, and so on – the forms which they draw or make, and which have shadows and reflections in water of their own are converted by them into images, but they are really seeking to behold the things themselves, which can only be seen with the eye of the mind. (Plato 1937, 510d-511a)

Kant held a similar view about the relation between genuine, ideal, "intuited" geometric figures and drawn, visible, "empirical" figures (Kant 1998, A74/B742-

A719/B747). Ever since Plato, most mathematicians, and even logicians, have taken the line that drawn diagrams and figures are merely heuristic devices, helpful aids, but not fit for rigorous demonstration. In the 17th Century, even Leibniz, who actually devised his own system for logic diagrams, held such devices to be secondary to formal proof. "The force of the demonstration is independent of the figure drawn, which is drawn only to facilitate the knowledge of our meaning" (Leibniz 1949, 403). As we will see, this attitude has only been seriously and extensively challenged by logicians during the past three decades. Yet there are still holdouts today. Not long ago, one prominent philosopher has ruled that a diagram "has no proper place in the proof as such" (Tennant 1986, 304). Here is how John Corcoran describes the situation:

> This anti-diagram view of deduction dominates modern mainstream logic. In modern mathematical folklore, it is illustrated by the many and oft-told jokes about mathematics professors who hide or erase blackboard illustrations they use as heuristic or mnemonic aids. (Corcoran 2009, 5, n11)

In spite of this, there has been of late, as I've said, a recent renaissance of interest focused on the powers of diagrammatic tools used in the tasks demanded of reason, especially deduction. A good measure of the depth and breadth of this new renaissance is the increasingly large number of logicians, philosophers, historians of logic and mathematics, mathematicians, and cognitive scientists who have been investigating a myriad of topics in the field of logic *diagrammatology* since the 1980s. One important reason for this has been the realization that a diagram, like any graphic device such as a map, bar graph, pie chart, etc., can be an efficient representation of significant amounts of information – and information is ultimately what logic is all about. So: "A diagram is a picture, in which one is intended to perform inference about the thing pictured, by mentally following around the parts of the diagram ... [It] is a store of information, from which inference can proceed" (Franklin 2000, 55). We can, and often do, reason with diagrams (whether visual or imagined). Of course, in doing so we are making use of certain logical principles. "The moral of the story is that when we reason with logic diagrams, we do not merely search for the best diagrams to *reason* with, we also look for the best logic to diagram with" (Moktefi 2015, 612). As we will see, the system of logic diagrams explored in this book is one that is partnered with a new system of term logic. The system is Term Functor Logic (TFL) and was pioneered by Fred Sommers.

Chapter II is devoted to highlighting key historical contributions to the development of diagraming in logic. It begins, where all serious study of logic begins, with Aristotle. Here, I argue than an examination of his *Prior Analytics* provides strong indications that Aristotle used some sort of diagramming system in pursuing and teaching his syllogistic system. A good case can be made that he used such diagrams to illustrate both syllogistic figure and mood. In the 17th and 18th centuries, first Leibniz, and then Lambert, devised a system for diagramming syllogistic inference using parallel but overlapping line segments. Soon after, Euler built a much more formidable diagram system for syllogistic using closed figures (viz., circles). In the 19th century, Venn, recognizing crucial limitation on the expressive powers of Euler circles, developed his own system of diagrams using closed figures. Venn diagrams became the inspiration for a number of other logicians to devise variations of Venn's original diagrams. By the end of that century, Peirce built a system of logic diagrams that had powers of expression and inference far beyond previous systems. At the same time, Frege offered a system of formal notation for what is now the standard first-order predicate calculus (Modern Predicate Logic). His system of notation, unlike today's standard systems, is two-dimensional, suggesting immediately that it could be construed as diagrammatic. Though post-Fregean logicians have tended to be generally hostile to the idea that any system of diagrams can play a substantial role in logical procedures, beginning in the 1980s, a growing number of logicians, as well as others, ushered in a new renaissance of diagrammatology.

Most of the work of those who have contributed to this renaissance has continued to rely on diagrams based primarily on closed figures. Chapter III begins with a brief critical examination of attempts at diagrammatic systems that, recalling those of Leibniz and Lambert, make use of line segments. The main aim of this chapter is to present my own system of linear diagrams. To that end, a short introduction to the basic elements of Sommers' Term Functor Logic is provided. Then, the various elements of the accompanying diagram system are given: the use of line segments to represent terms, points to represent singular terms, vector arrows to represent relational terms, and a way of diagrammatically representing unanalyzed (into terms) sentences (whether simple or compound). Along the way, methods for using such diagrams in the procedures of inference are explained and illustrated.

Chapter IV begins by examining some important work by logicians, philosophers, and cognitive scientists studying the powers and limits of logic diagrams. Much of this work has made use of empirical studies that reveal both some of these powers and some of these limits. Linear diagrams also have their powers and limitations. The system offered here has been shown to be limited in its expressive power by constraints imposed on any geometric figure (whether open or closed). After considering this, a response is then provided. Next, a survey of attempts by cognitive scientists to determine just how ordinary people, untrained in any formal system of logic, actually carry out various tasks of reasoning. One of the things revealed here is that researchers are divided over the question of whether such reasoning depends on a prior, perhaps innate, reasoning capacity or, instead, reasoning is carried out by the manipulation of "mental models" (internal diagrams). Finally, some studies suggest that ordinary reasoners are more likely to make use of such internal diagrams than rules for manipulating sentential expressions.

Chapter V is provided by José Martin Castro-Manzano and José Roberto Pacheco-Montes. Here they do two important things: first, expand the scope of TFL by showing how to incorporate quantifiers ('many', 'most', 'few') into the formal system; second, augment our linear diagram system (ED) so that expressions (and thus inferences) involving such non-classical quantifiers can be represented as well.

Finally, Chapter VI summarizes the main features and lessons found in the preceding chapters. The conclusion that the system of diagrams offered here is both practice-based and practical is reinforced.

In the end, a relatively extensive Bibliography and Further Readings is provided in order, in part, to invite readers to explore more fully the general field of diagrammatology in logic.

2 Some Historical Figures

> It is not unlikely that [Aristotle] represented each figure of the syllogism by a different geometrical figure, in which the lines stood for propositions and the points for terms.
>
> W.D. Ross

> I think no one who understands the matters doubts that the part of logic which deals with the moods and figures of the syllogism can be reduced to geometrical rigour.
>
> Leibniz

> Don't think you can outsmart Aristotle!
>
> Krister Segerberg

2.1 Aristotle and the Lost Diagrams

The idea that diagrams are merely aids to the intellect, training wheels for the novice, nothing more than heuristic devices is ancient and persistent. Plato certainly had a hand in this. Consider this from *The Republic:*

> Don't you also know that they [mathematicians] use visible forms besides and make their arguments about them, not thinking about them but about those others that they are like? They make the arguments for the sake of the square itself and the diagonal itself, not for the sake of the diagonal they draw, and likewise with the rest. These things themselves that they mold and draw, of which there are shadows and images in water, they now use as images, seeking to see those things themselves, that one can see in no other way than in thought. (Plato 1968, 191)

Yet, has any serious reader of Plato's *Meno* ever read the slave boy sections without drawing (or wanting to draw) a series of diagrams matching the instructions Socrates draws from the boy's responses to his questions? Ancient Greek mathematicians seem to have invented the notion of logical proof, thus making mathematics (and, after Aristotle, logic) a coherent body of rational knowledge rather than just a collection of mathematical observations. They used the word δεικνυμι ('deiknumi'), 'show', 'display', 'illustrate', 'point out', 'make visible', for *proof* (see Szabó 1978, 286 ff). Centuries before the Pythagoreans, it was well known that, at least for a right plane triangle whose sides are in the ratio 3, 4, 5, the square of the hypotenuse equals the sum of the squares of the other two sides.

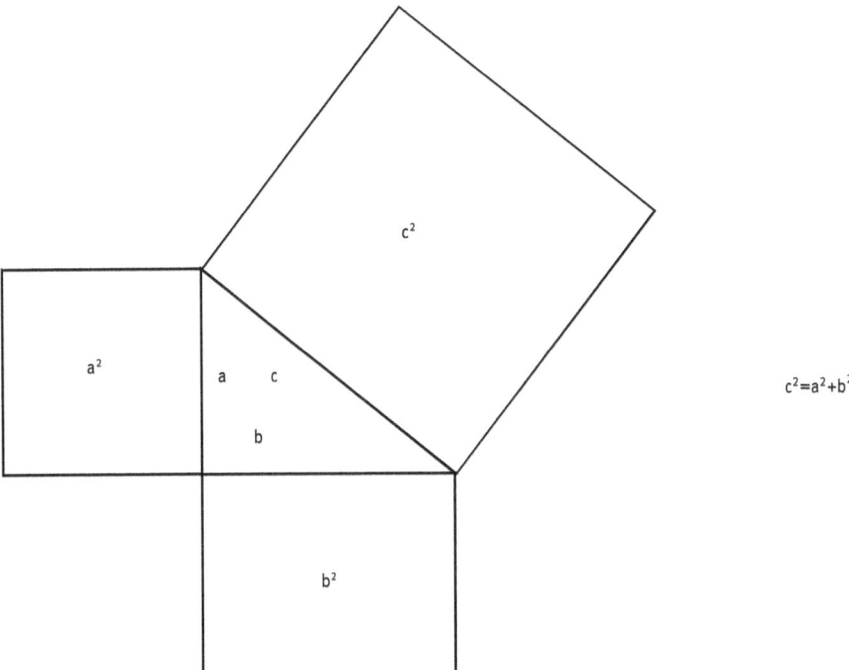

Figure 2.1: The Pythagorean Theorem

In the 6th Century bce, the Pythagoreans demonstrated that this holds in general for *any* right triangle. In doing so, their proof linked geometry with algebra (or at least arithmetic) and both to physical space. And they did the same with the Golden Ratio, which characterized the *pentacle*.

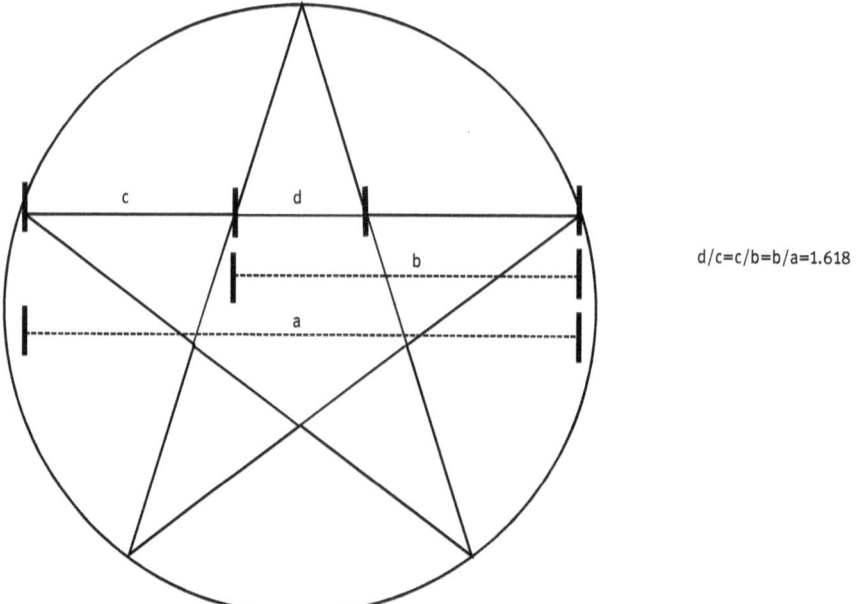

Figure 2.2: The Pentacle and the Golden Ratio

In such cases there is no doubt that geometry came before algebra, diagrams before calculations. This needn't have been the case. One could imagine an ancient mathematician conceiving of the formula '$a^2+b^2=c^2$' and then seeing if it applied to any geometric figure. But eventually ancient mathematicians faced a crisis: irrational numbers. Such numbers cannot be expressed as fractions whose numerator and denominator are whole numbers (e.g., 1/3); nor (much later with the development of decimal notation) could they be expressed as non-repeating infinite decimals (e.g., .33333...). Only approximation was possible when it came to the expression of irrational numbers. On the other hand, they can often be clearly and economically displayed diagrammatically. The ratio (called π, 'pi') between the circumference and the diameter of a circle is easily illustrated but, being an irrational number, can only be numerically expressed approximately. Indeed, ancient mathematicians based their approximations on geometric figures, inscribing a series of regular polygons, of ever increasing numbers of sides, in a given circle.

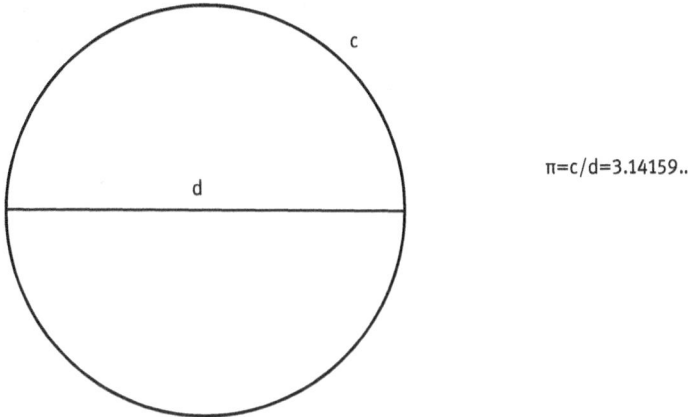

Figure 2.3: π

The square of the diagonal of a unit square is, by virtue of the Pythagorean Theorem, 2. Thus, the length of the diagonal must be the square root of 2, √2, which, is another irrational number, easily illustrated by a figure but not easily or completely expressed numerically.

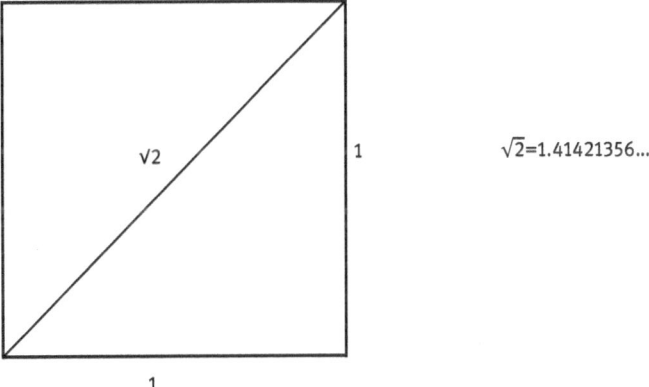

Figure 2.4: The Diagonal of a Unit Square

It is reasonable to conclude that for ancient mathematicians (before the discovery of irrational numbers) geometry and algebra (or at least arithmetic) could be seen as mutually supporting each other. Their geometry alone led them to the irrationals. Their "geometric knowledge" seemed to go beyond their mathematics. Their figures allowed them to see what their mathematics could not show.

While the well-known tradition of claiming that above the portals of the Academy Plato had inscribed, "Let none untrained in geometry enter here" may be just a myth, there is no doubt that geometry played an important role in the Academy. Having spent two decades there, Aristotle was certainly not ἀγεωμέτρητος, "untrained in geometry." He was aware of the methods of geometric demonstration that were, and had been, employed by mathematicians. Euclid's was far from the first attempt to lay out the "elements" of geometry. Like his teacher, Aristotle saw diagrams as valuable aids to the understanding (especially for the student) but nothing more. "[W]e (I mean the student) use the setting out of terms as one uses sense-perception; we do not use them as though demonstration were impossible without these illustrations, as it would be without the premisses of a syllogism." (*Pr. An.* 49b40–50a4) Mention of, or allusions to, the mathematicians' use of geometrical diagrams are common throughout Aristotle's work. For example, in *Metaphysics*, after offering a diagrammatic demonstration that the angles of a triangle equal two right angles, he wrote, "It is by an activity also that geometrical constructions are discovered; for we find them by dividing" (1051a22, Ross translation), adding that "it is by making constructions that people come to know them" (1051a32, Ross translation). But in *Prior Analytics* he is talking about syllogisms and *logical* demonstration. More importantly, it seems obvious that Aristotle, at least as a teacher, was making use of diagrams, "setting out" the terms of a syllogism so that they can be *seen* ("as one uses sense-perception"). Consider the following passages from *Prior Analytics* (unless noted otherwise, all translations from *Prior Analytics* are taken from R. Smith:

> Whenever, then, three terms are so related to each other that the last is in the middle as a whole and the middle is either in or not in the first as a whole, it is necessary for there to be a complete deduction of the extremes. (I call that the *middle* which both is itself in another and has another in it – this is also the middle in position – and call both that which is itself in another and that which has another in it *extremes*.) (25b32–38)
>
> If one term belongs to all and another to none of the same thing, or if they both belong to all or none of it, I call such a figure the *third*. By the *middle* in it I mean that term of which they are both predicated, and by *extremes* the things predicated: by *major* extreme I mean the one further from the middle and by *minor* the one closer. The middle is placed outside the extremes and is last in position. (28a10–15)

Here Aristotle makes use not only of (Greek versions of) such terms as 'figure', 'extreme', and 'middle' but 'farther from', 'closer', 'placed outside', and 'last in position'. The notion of containment (being in) also appears to play a prominent role here. Drawing from these clues, the conclusion that Aristotle was in the habit of teaching and illustrating syllogistic inference by including the use of some kinds of diagrams. Concerning the passage quoted above from *Pr. An.*, 25b31–38, Geach noted, "Clearly a reference to a diagram, now lost" (Geach 1987, 27, n. 3). As W.D. Ross wrote, "It is not unlikely that [Aristotle] represented each figure of the syllogism by a different geometrical figure, in which the lines stood for propositions and the points for terms" (Ross 1960, 37). Indeed, Carlo Natali has argued that Aristotle made extensive use during his lectures of diagrams, tables, maps, charts, models, drawings, etc. (Natali 2013, 113–117). Among the textual passages cited by Natali are the following: "For the sake of example, let's take each of them and study them from the diagram ..." (*Eud. Eth.*,1220b36). "Let's observe in the following diagram what we are talking about ..." (*De Int.*, 22a22). "Let the outer rainbow be the [line] B, and the inner and primary one be the A; as for the colours, let the C be red, the D be green, and E be violet, and the yellow appears at the [point] F." (*Meteo.*, 375b9–12).

We will shortly return to Aristotle's practice of constructing and using diagrams in logic. There we will see the import of Ross' remark quoted above. For now, let us consider what is unquestionably the most famous diagram in logic, the Square of Opposition, the origins of which are to be found in Aristotle's *De Interpretatione*. The *diagrammed* square of opposition appeared early in the common era in commentaries on Aristotle's logic (particularly *De Interpretatione*). Such diagrams were common during the high Middle Ages and continued to be found in logical works thereafter. There is no textual evidence that Aristotle actually produced a diagrammed square of opposition but he did, in effect, describe such a square and gave clear examples of sets of four categorical statements that satisfy the logical relations that the square is meant to exhibit. He even placed such sets in a square-like pattern. In *De Interpretatione*, chapters VI, VII, X, he drew distinctions between various kinds of logical opposition. "Each affirmative statement will have its own opposite negative, just as each negative statement will have its affirmative opposite. Every such pair of propositions we, therefore, shall call contradictories ..." (*De Int.* 17a33–35, unless specified otherwise, all translations from *De Interpretaione* are by Tredennick). Given that statements can be universal or particular quantity, Aristotle claimed that a pair of statements, "one affirmative, one of them negative, both [quantitatively] universal in form, having one universal [i.e., general term] for subject; then these propositions are contrary." (*De Int.* 17B3–6). Indeed, he went on to say that such contrariety between given pairs of statement is due to the contrariety that holds between the qualities being predicated. "While two propositions that are true can together be truly asserted, two contrary propositions must predicate contrary qualities, and these in the selfsame subject can never together inhere." (*De Int.* 24a8–9) As well, a pair of statements that share a common subject but are not universal in quantity are also somehow logically opposite in a way that might be considered contrariety (*De Int.* 17a7–10). The three kinds of logical oppositions Aristotle was accounting for here are *contradiction, contrariety,* and what came to be called *sub-contrariety.* He characterized them both syntactically and semantically. In Ackrill's translation (Ackrill 1987, 15):

> I call an affirmation and a negation *contradictory* opposites when what one signifies universally, e.g. 'every man is white' and 'not every man is white', 'no man is white' and 'some man is white'. But I call the universal affirmation and the universal negation contrary opposites, e.g. 'every man is just' and 'no man is just'. So these cannot both be true together, but their opposites may both be true with respect to the same thing, e.g. 'not every man is white' and 'some man is white'. Of contradictory statements about a universal taken universally it is necessary for one or the other to be true or false; similarly, if they are about particulars, e.g. 'Socrates is white' and 'Socrates is not white'. But if they are about a universal not taken universally it is not always the case that one is true and the other false. For it is true to say at the same time that a man is white and that a man is not white, or that a man is noble and a man is not noble .... (*De Int.* 17b16–33)

Aristotle distinguished between denial of a predicate to a subject and the negation of a term, for example 'not-man'. He called that latter "indefinite nouns" (*De Int.* 16a30–33). All these distinctions (among kinds of logical opposition and kinds of negation) are preparation for Aristotle's analysis of the logical relations found displayed on the square of opposition. At *De Int.*19b24–30, he set out what he called "the four" as follows:

> Supposing, I mean, the verb 'is' to be added to 'just' or 'not-just', we shall have two affirmative judgements; supposing that 'is not' is added, we then have two negative judgements. Together these make up the four. This the subjoined examples make clear: –
>
> Affirmations   Negations
> [Some] man is just   [Some] man is not just
> [Some] man is not-just   [Some] man is not not-just

Aristotle then went on to show that the same arrangement holds when the statements are universally quantified:

Every man is just   Not every man is just
Not every man is not just   Every man is not just

Not quite the familiar square of opposition. However, keeping in mind that for Aristotle there was no such operation as sentential negation, the contradictory of a statement is not its negation; it is the denial of its *predicate* (not to be confused with the affirmation of the negation of the *predicate term:* '... is not P' vs '...

is not-P'). So it is possible to arrange Aristotle's two sets of four statements in such a way that they fit nicely on the square:

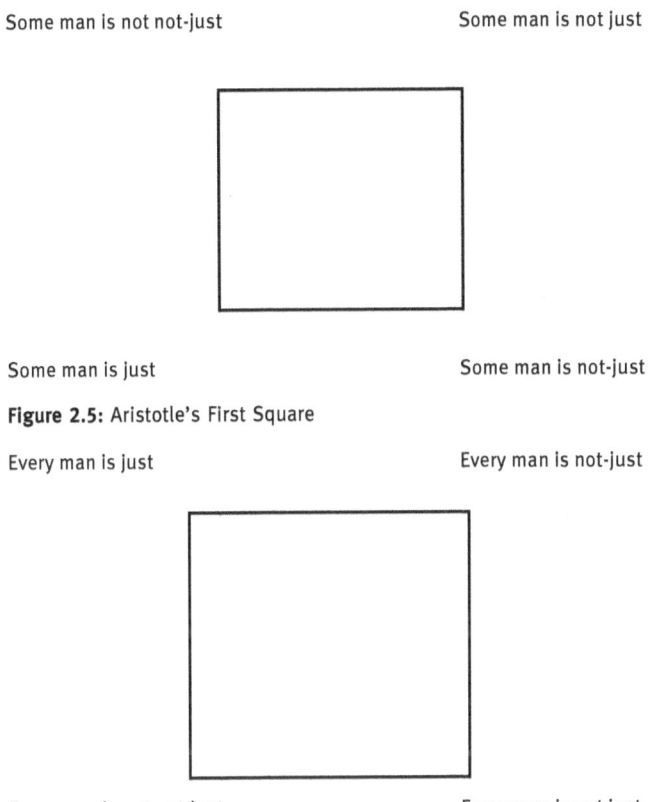

Some man is not not-just   Some man is not just

Some man is just   Some man is not-just

Figure 2.5: Aristotle's First Square

Every man is just   Every man is not-just

Every man is not not-just   Every man is not just

Figure 2.6: Aristotle's Second Square

Squares of opposition are used to illustrate the logical relations of contradictoriness, contrariety, sub-contrariety, and (usually) subalternation. The origin of such squares is found in Aristotle. However, what one generally has in mind in considering logical diagrams are visual representations of deductions (especially syllogistic deductions). And there is no doubt that the origin of not only formal logic, particularly syllogistic, but even more particularly such diagrams, is likewise found in Aristotle.

In *Prior Analytics*, Aristotle adhered to a theory of logical syntax that analyzed all statement-making sentences (statements) as consisting of a pair of terms bound together to form a unified sentence (rather than merely a sting of words) by means of a *logical copula* (English versions of which are 'belongs to all', 'belongs to no', 'belongs to some', and 'does not belong to all'). This theory contrasts with one that he had probably held before he took on the project of building a formal logic in *Prior Analytics*, one he no doubt learned from Plato. Plato took a sentence to consist of a string of two terms, one of which must be a noun and the other of which must be a verb. In the *Sophist*, he wrote that a sentence cannot be formed by a pair of nouns, nor by a pair of verbs (261d). Thus, "the first and smallest" sentence is a combination of a noun and a verb, a combination resulting from the two being "mingled" (262c). According to Plato's theory of logical syntax, a sentence consists of just two expression, while on Aristotle's it consists of three (two terms and a connecting copula. In *Prior Analytics*, Plato's *binary* account (which turns out to be shared in its essentials by post-Fregean logicians) gives way to Aristotle's *ternary* account (for much more on this see, for example Sommers 1976a, 1976b, 1976c, 1976d, 1982, 1983, 1990, and Englebretsen 1982b, 1996, 1997, 2002). Aristotle's new theory of logical syntax turned out to be a necessary prerequisite for the logic of syllogisms. A two-premise syllogism requires at least one of its three terms to appear as a subject term in one statement and as a predicate term in another. Terms, now undifferentiated into nouns and verbs, are fit to do just that (*Pr. An.* 43a25–44).

As we have seen, when articulating his syllogistic system in *Prior Analytics*, Aristotle made extensive use of spatial vocabulary. Consider once more the passage quoted earlier (*Pr. An.* 25b32–38):

> Whenever, then, three terms are so related to each other that the last is in the middle as a whole and the middle is either in or not in the first as a whole, it is necessary for there to be a complete deduction of the extremes. (I call that the *middle* which both is itself in another and has another in it – this is also the middleposition – and call both that which is itself in another and that which has another in it *extremes*.

We saw that both Ross and Geach took such passages as providing strong evidence that Aristotle was making use of syllogistic diagrams (using, as Ross suggested, lines for propositions and points for terms). Suppose Aristotle was in the habit of teaching and illustrating syllogistic deductions including the use of diagrams consisting of points and line segments (things easily drawn on any handy surface). If so, then the passage above could be read (as they have been by most) as descriptions of Barbara and Celarent syllogisms diagrammed as follows (with S as minor, M as middle, and P as major terms):

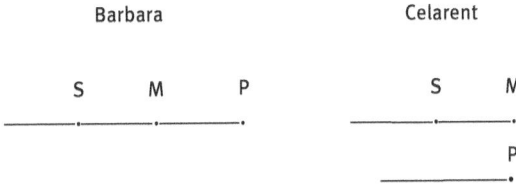

Figure 2.7: Possible Line Diagrams for Barbara and Celarent

Notice that now the relations of being wholly contained in and wholly excluded from are simply the containment of one line segment in another (larger) line segment and the exclusion of one line segment from another, where line segments are marked at their right terminal points. The middle term (point) is literally in the middle. The deduction is complete in the sense that the diagram for each conclusion can be *seen* once the two pairs of premises have been diagrammed.

What sorts of syllogistic diagrams did Aristotle use? Were they like the ones illustrated above? What kinds of geometric elements would they have involved? Were letters, words, or other expressions part of a diagram?

Let us begin with Aristotle's syllogistic *figures* (σχηματα, schemata) a term that on its own suggest something with a shape. As we have seen, a statement was taken to be logically parsed as a pair of (grammatically undifferentiated) terms bound together by one of Aristotle's four logical copulae. The figure of a syllogism depends on the orders of each pair of terms in each of the premises of the syllogism. Omitting for now the copula, we can think of the ordered pair of terms (predicate term followed by subject term) as a *proto-statement*. Following Aristotle's normal practice, we can use uppercase Greek letters as variables standing for terms. For example, 'R belongs to all M' could formulate 'Rational(ity) belongs to all men'. The proto-statement form of such a statement would then be 'RM'. It will turn out that any number of statements will share a common logical form; and at least four such forms will share a common proto-statement form. For example, the formulae 'A belongs to all B'. 'A belongs to no B', 'A belongs to some B', and 'A does not belong to all B' all share the proto-sentential form 'AB'. Again, the figure of a syllogism depends on the proto-statement forms of the two premises, with the order of the premises as follows: the first premise is the one that contains the term which is the predicate term of the conclusion, the second premise is the one that contains the term which is the subject term of the conclusion, and the remaining term is contained in each premise but not the conclusion. The first premise has traditionally been called the *major premise*, the second premise is the *minor premise*, the predicate term of the conclusion is the *major term*, the subject term of the conclusion is the *minor term*, and the term contained in the two premises but not the conclusion is the *middle term*. A syllogism such as:

> Some men are logicians.
> Every man is rational.
> Thus, some rational (things) are logicians.

can be paraphrased, using the appropriate Aristotelian copulae as:

> Logician belongs to some men.
> Rational(ity) belongs to all men.
> Thus, logician belongs to some rational (things).

This can then be formulated as:

> L belongs to some M
> R belongs to all M
> Thus: L belongs to some R

The figure of this syllogism is then simply determined by the proto-statement forms of its premises:

> LM
> RM

A bit of thought reveals that once the order of the premises is settled (major, minor) the figure of any syllogism is then just a matter of the position of the middle term. For Aristotle, the middle term can occur in one of three ways in the premises of a syllogism: (α) it can be the predicate term of both premises, (β) it can be the subject term of both premises, or (γ) it can be the predicate term of one premise and the subject term of the other. Let P, M, and S stand for the major, middle, and minor terms of a syllogism. Then the three possibilities are:

> (α)   (β)   (γ)
> MP   PM   PM MP
> MS   SM   MS SM

As it happens, Aristotle took the two (γ) arrangements as constituting a single syllogistic figure (the *first figure*), the (α) arrangement as the *second figure*, and the (β) arrangement as the *third figure* (these are described in *Prior Analytics*, chapters IV, V, and VI, respectively). Later logicians took only the first of the two (γ) arrangements as the first figure and called the other version the *fourth figure*. Given that the tacit conclusion in every case has the proto-statement form: PS, the traditional four syllogistic figures then are:

First Figure	Second Figure	Third Figure	Fourth Figure
PM	MP	PM	MP
MS	MS	SM	SM

Needless to say, though every two-premise syllogism fits one of the four figures, figure alone is not enough. What Aristotle was aiming for (among other things) was a way of distinguishing between *valid* and *invalid* syllogisms. Since any number of statements can share a given proto-statement form, what distinguishes them from one another, from the perspective of logical syntax, is the difference among their copulae. Consider the fact that there are four kinds of copluae (thus four kinds of categorical statements that can serve as a premise or a conclusion), and that three such statements comprise a syllogism. This means that there are sixty-four possible combinations (traditionally called *moods*) of three statements. Since there are four possible configurations (figures) for each possible combination, there are 256 possible syllogistic forms. Aristotle took nineteen of these to be valid. An additional five, so-called *weakened* forms were also viewed as valid. For Aristotle, every argument is either valid or invalid. An argument is valid just in case its conclusion is a logical consequence of its premises. It is an argument that cannot possibly have a false conclusion if its premises are all true. It is just such a valid argument that Aristotle calls a *syllogism* (συλλογισμὸν). According to Tredennick's translation (*Pr. An.* 24b18–22):

> A syllogism is a form of words in which, when certain assumptions are made, something other than what has been assumed necessarily follows from the fact that the assumptions are such. By 'from the fact that they are such' I mean that it is because of them that the conclusion follows; and by this I mean that there is no need of any further term to render the conclusion necessary.

An invalid argument, one whose conclusion is not a logical consequence of its premises (thus not officially a syllogism at all) can have all true premises but a false conclusion. There are two kinds of valid arguments and two kinds of invalid arguments. A valid argument, a syllogism, is either one that can be *seen* to be (is intuitively) valid or one that can be *shown* to be valid by means of a *deduction*. Aristotle took the first of these syllogisms to be *perfect* while the second ones were *imperfect*. He wrote (*Pr. An.* 24b23–27, Tredennick translation):

> I call a syllogism perfect if it requires nothing, apart from what is comprised in it, to make the necessary conclusion apparent; imperfect if it requires one or more propositions which, although they necessarily follow from the terms which have been laid down, are not comprised in the premises.

A deduction is, in effect, a chain of statement forms beginning with the premises and ending with the conclusion, such that every link in the chain is the result of using rules (where a rule is either a license for reformulating (*converting*) a line or a perfect syllogism). Aristotle also allowed a second type of deduction (*Pr. An.* 45b1–5). Since it is impossible for any valid argument (syllogism) to have all true premises and a false conclusion, an argument consisting of the premises and the contradictory of the conclusion must logically lead to a contradiction. So a deductive chain that begins with the premises and the contradictory of the conclusion and ends with a statement that is the contradictory of a previous statement in the deduction is an *indirect* deduction (what became known as a "proof *per impossible*"). In the case of an invalid argument, either its premises are *known* to be all true and its conclusion is *known* to be false, or, if these things are not known, it can be shown to be invalid by a *counter-argument*, another argument that does have premises that are all known to be true and a conclusion that is known to be false – and *both arguments share the same logical form*. It is Aristotle's emphasis on logical form that gives his syllogistic system the universal applicability, and much of the power, that it has. An argument that is valid has a valid/invalid form; *any* argument sharing that form is also valid/invalid. His is a (indeed, the first) system of *formal logic*. (One could add as well that, given his use of letters as variables for terms, his is also the first *symbolic* logic.)

Aristotle's overall goal in *Prior Analytics* was to provide a theory of logical deduction, which was particularly important for the goal he set out for his *Posterior Analytics*, namely, a theory of logical *demonstration*.

> Having made these determinations, let us now say through what premises, when, and how every deduction [of an imperfect syllogism] comes about. (We will need to discuss demonstration later. Deduction should be discussed before demonstration because deduction is more universal: a demonstration is a kind of deduction, but not every deduction is a demonstration. (*Pr. An.* 25b26–31)

Demonstration is required for syllogisms whose premises are true so that the conclusion must be true. This is required for sciences like physics, where knowledge is built up from given truths to new truths. Deduction does not demand that the syllogism's constituent premises and conclusion be true, only that the premises cannot all be true while the conclusion is false. That is why deduction (the theory of perfect syllogisms and imperfect, but deducible, syllogisms), is "more universal". Ultimately, a demonstration rests on axioms, truths that require no deduction. Physics may be an *axiomatic* system of science, but syllogistic is not (it is a *natural deduction* system, as Smiley 1973 and Corcoran 1974 have shown). As it happens, not only is every deduced syllogism valid, but every valid syllogism can be deduced. The syllogistic system of logic is sound and complete (see especially Corcoran 1972 and Smiley 1994).

It's time now to return to the questions posed earlier. What sorts of diagrams did Aristotle use? Were they like the ones illustrated above (Fig. 2.7)? What kinds of geometric elements would they have involved? Were letters, words, or other expressions part of a diagram?

The best recent work done in trying to answer some of these questions has been by Marion Wesoły (see Wesoły 1996 and 2012, but see also Einarson 1936, 165–169). The starting point here is an understanding of just what Aristotle meant by *analytics* in *Prior Analytics*. Aristotle understood his task to provide both an analysis of premise(s)-conclusion arguments and an account of deduction (primarily reduction to perfect syllogisms).

> For Aristotle, to 'analyze' is to make an investigation of how to discover terms and premises from which to deduce the desired conclusion, or of how to resolve a conclusion into its terms constructing premises. Thus, the Greek geometrical analysis and Aristotle's analytics constitute a heuristic or regressive procedure: from a given *problema* (conclusion) to grasp, by means of diagrams, the relevant elements as terms and premises of the syllogism. Conversely, to this heuristic procedure, *syllogismos* constitutes a *synthesis*, namely a deductive or progressive reasoning. ... Much has been written on Aristotle's syllogistic as a deductive procedure, but its analytical or heuristic strategy of finding terms and premises of syllogisms has generally been overlooked, mainly because the relevant diagrams of the three figures have not been adequately taken into account. (Wesoły 2012, 89–90)

Analytics provides an account of an argument in terms of its origins, its genesis, its elements. Ultimately, the elements are terms. Analytics begins with a statement (conclusion) and then seeks other statements (premises) whose acceptance can necessitate that statement (see Wesoły 2012, 85–86). The key to Aristotelian logical analysis is the discovery of a middle term, a term that is shared by a pair of premises but not the conclusion. Such a term can appear in only three ways in the premises: as the predicate term of each, as the subject term of each, or as the predicate term of one and the subject term of the other. These three different positions for the middle term determine the three Aristotlesian syllogistic figures. σχήματτα. Wesoły's aim is "to recapture, at least to some extent, Aristotle's probable diagrams, which regrettably are missing from the extant text of the *Prior Analystics*" (Wesoły 2012, 83–84). The diagrams that he reconstructs are, in effect, diagrams of the figures.

Aristotle defined a syllogism at *Prior Analytics* 24b18, and soon after gave descriptions of each of the three syllogistic figures 25b32–28a18 and 47b1–9, using such expressions (translated as) 'middle', 'extreme', 'lying next to the middle', 'outside the extremes', 'first from the middle', 'lying next to the middle', 'farther from the middle', 'closer', and 'last in position'. Moreover, as Wesoły points out, Aristotle did not expressly use the notion of *premise*, but instead he used 'protasis'. This refers to something *put forward*, and, when used in talking about terms, "'stretching' as of lines joining two points" (Wesoły 1996, 58 and 2012, 90). Finally, at 46a9, Aristotle explicitly used the word διαγεγραμμένων ... diagrammed. In light of all this and more, it is hard to deny that Aristotle illustrated his logic with diagrams. "Hence, the invention of the syllogistic figures can be seen as an extension of the diagrammatic analysis" (Wesoły 2012, 89). "Strictly speaking, the notion of analytical syllogism make sense only if we interpret it within the framework of the three figures, which implicitly refer to a syntactical position of terms in diagrams that are now lost, but that once accompanied the text of the *Prior* Analytics (Wesoły 2012, 92).

As we have already seen, the mere proto-statement forms of the premises are sufficient to determine the figure of a standard two-premise syllogism. Thus, in attempting a reconstruction of Aristotle's diagrams for the figures, Wesoły's primary concern is with proto-statement forms, with the forms of full categorical statements (those displaying the appropriate copulae) of secondary concern. His practice, like Aristotle's, is to use uppercase Greek letters as term variables (with A, B, and Γ as the major, middle, and minor terms respectively). While syllogistic premises and conclusions are usually presented linearly, one-dimensionally (e.g., AB) as in most written texts, including Aristotle's, Wesoły's view is "that in Aristotle they were originally two-dimensional" (Wesoły 2012, 92). He notes that at *Posterior Analytics* 88a34–36 Aristotle wrote of attaching "either into the middle [terms] or from above or from below, or else to have some of their terms inside and others outside" (Wesoły 2012, 92, n. 25). This vocabulary of expressions such as 'into the middle', 'from above/below', etc., indicates that not only had Aristotle been making use of some system of two-dimensional diagrams but that such diagrams were somehow vertically oriented. Displaying predicate terms above subject terms and using arrows to indicate the direction of predication, the proto-statement form AB would be diagrammed as:

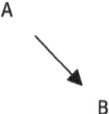

Figure 2.8: Diagram of AB

Wesoły then uses double arrows to indicate the direction of the predication for conclusions. So he offers the following diagrams for the three Aristotelian figures (first, second, third, respectively):

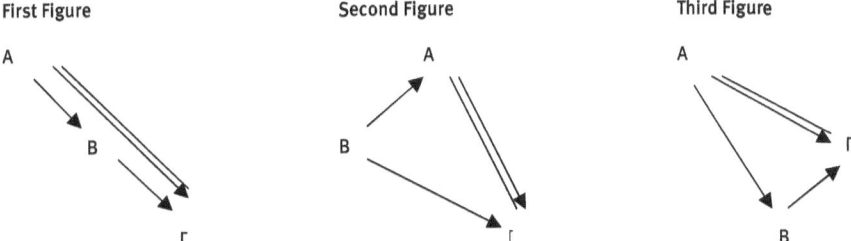

Figure 2.9: Diagrams for the Three Aristotelian Figures

So far, no indication of the predication relations, copulae (indicating both the quantity and quality of each syllogistic statement), has been indicated. Wesoły simply attaches the appropriate medieval letter (*a, e, i, o*) to each arrow. But these only *say* what the logical (categorical) form of each premise and the conclusion are; they do not *show* what they are. And surely showing is the point of diagramming. At any rate, an example of such a marked diagram would be this one for Cesare:

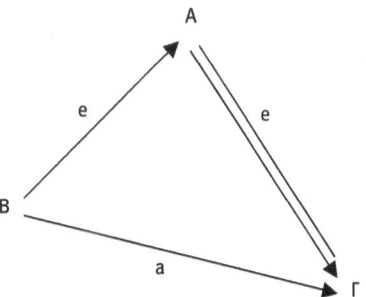

Figure 2.10: Diagram for Cesare

Notice that in such diagrams there is both a vertical axis (along which the major, middle, and minor terms are arranged, top to bottom) and a horizontal axis (along which the middle is arranged, left to right, with the predicate terms always to the left of their subject terms). Having established these diagrammatic conventions as standard, Wesoły is able to drop the arrow notation all together, while displaying the conclusion horizontally below a line separating it from the premises in such a way that both tokens of each extreme term are vertically aligned (Wesoły 2012, 99–100). The following pair of examples illustrate the results:

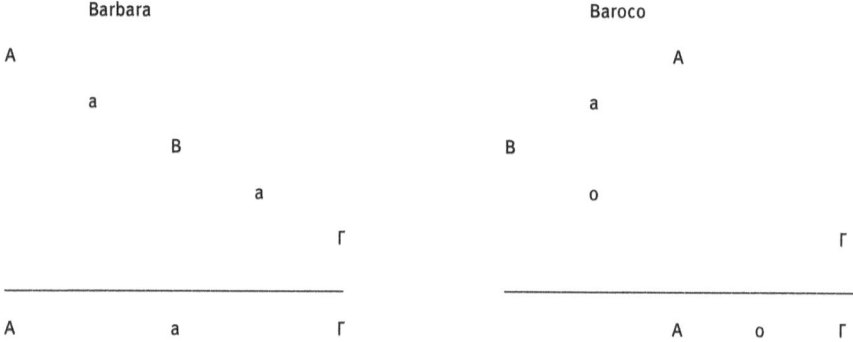

Figure 2.11: Barbara and Baroco

According to Wesoły, "It seems likely that Aristotle employed such configurations of blackboard diagrams" (Wesoły 2012, 99). This might well be the case. However, such diagrams, while being able to visually display the *analytic* direction of syllogistic based on the schematic figures (and to be fair, this is just what Wesoły set out to do), they do not yet display the *synthetic* direction of syllogistic. As we know, a genuine syllogism for Aristotle at least, was a valid argument. A perfect syllogism "requires nothing apart from what is comprised in it to make the necessary conclusion apparent," while an imperfect (but valid) syllogism "requires one or more propositions ... not comprised in the premises" (*Pr. An.* 24b23–27). Analysis begins with a purported conclusion and then seeks other statements (premises) such that it is "because of them that the conclusion follows" (*Pr. An.* 24b22). The appropriate premises are found by the discovery of a term (the middle term) that connects the premises to one another and, thereby, connects the two extreme terms to one another. Synthesis, by contrast, begins with an imperfect syllogism and then seeks to show how, beginning with the premises, a series of linked implications lead eventually to the conclusion. In effect, this means "completing" ("reducing") a given imperfect syllogism to one of the four perfect syllogisms of the first figure by means of rules of such reduction: conversion, *reduction ad impossible*, and *ekthesis*. As we saw earlier, Wesoły understands the analytic and synthetic procedures "constitute a complementary and convertible order of inquiry" (Wesoły 2012, 90).

An ideal system of diagrams for Aristotle's syllogistic logic would include ways to exhibit the deductions involved in reducing imperfect to perfect syllogism. This, in turn, would require a means of systematically displaying the predication relations (indicated thus far by the *a, e, i, o,* notation) visually. In other words, a way of exhibiting *moods* as well as *figures* is required. Wesoły seems to suggest (Wesoły 2012, 102–105) that these predication relations (what he calls "categorical predications") are to be understood as resulting from the semantics of Aristotle's theory of categories found in his *Categories*, and upon which the analysis of the syllogistic figures rests (Wesoły 2012, 105). There is no doubt that Aristotle's understanding of the categories had an influence on his formal logic. An important example of this is his recognition of inferences involving relational terms, examples of which are found at *Prior Analytics* 48b15–27 and 48b31–39 (for an analysis see Englebretsen 1982a). While Wesoły's work on reconstructing Aristotle's lost diagrams of the syllogistic figures will not be the last word on how to diagram syllogisms, it is certainly the best work done so far. Importantly, it profits from a recognition that "The most remarkable feature of Aristotle's analytics as a whole is undoubtedly its ingenuity, thoroughness and perspicacity" (Wesoły 2012, 109). He adds:

> Contemporary logicians seem to have no patience and cognitive curiosity for Aristotle's analytics, as they rashly neglect or belittle the importance of his formulation of the analytical figures. The point of view of modern mathematical formal logic is obviously instructive and illuminating, but ... it is quite differently-oriented, and, therefore, it may sometimes prevent us from obtaining a historically adequate interpretation of Aristotle's achievements. (Wesoły 2012, 109)

And one more thing about investigating systems of logical diagrams:

> Nowadays, when we investigate the syllogistic and the diagrams, we come to think of yet [other] famous proposals of the diagrams, i.e., the ones by Euler and Venn. However, these accounts are quite distant from the ancient and medieval ones, as they concern the tracing of syllogistic validity, but no longer have any strict connection with Aristotle's analytics and the relevant reduction to the figures. For this reason we omit them here. (Wesoły 2012, 107)

2.2 "The Golden Age" – Leibniz, Lambert, and Euler

[T]he best preparation for original work on any philosophical problem is to study the solutions which have been proposed for it by men of genius whose views differ from each other as much as possible.

C.D. Broad

[T]the circles, or rather these spaces, for it is of no importance of what figure they are of, are extremely commodious for facilitating our reflection on this subject, ... by means of these signs the whole is rendered sensible to the eye.

Euler

That long period between the time when Aristotle began the systematic inquiry into formal reasoning, and along with it the process of devising a workable (but now lost) system for displaying such reasoning diagrammatically, and the much later period when others developed systems of logical diagrams that we find familiar today, was not completely devoid of interest in logical diagrams. But the fact is that, aside from the extensive use of diagrams as pedagogical aids, such as squares of opposition (used to display the logical relations that hold among the four categorical forms of statements) and the *pons asinorum* (used to find possible middle terms when given a conclusion), actual logical diagrams, the kind that can be used to "carry out logical reasoning" and "solve logical problems" (Moktefi and Shin 2012, 611–612), seem not to have been used at all. The development of systems of such diagrams finally began in what Moktefi and Shin call "The Golden Age of Logic Diagrams" (Moktefi and Shin 2012, 616), the age of Leibniz, Lambert, and Euler. It was a period when logicians had no doubt that Aristotle's syllogistic system, perhaps with an adjustment here and an emendation there, would be the final word on logical reckoning. In fact, not much later, Kant felt justified in claiming that logic (Aristotle's logic) was, for all intents and purposes, finished and complete. The golden age of logic diagrams was, to be precise, a golden age of diagrams for syllogisms.

As we have seen, Wesoły made clear the distinction between syllogistic analysis and syllogistic synthesis. He reconstructed plausible versions of Aristotle's lost diagrams for the former. These made use of vector lines with endpoints labelled by term letters. We have also seen that an ideal system of logical diagrams must display syllogistic synthesis as well. It must be able to illustrate visually just how conclusions are derived from premises. Here Wesoły's labelling of vectors with the a, e, i, and o notation is not sufficient. Earlier we asked whether Aristotle, in teaching and illustrating syllogistic deductions, might have used diagrams such as those suggested in Figure 7 (Possible Line Diagrams for Barbara and Celarent). He almost certainly used a stylus and waxed tablet in writing.

However, in a pedagogical setting, when instructing a number of students, it is more likely that he drew whatever diagrams he needed on a large surface (probably on the ground, in dirt or sand). In such cases, the easiest and simplest kinds of diagrams would consist, initially at least, of points and line segments. Moreover, he could have avoided the need to visually label these in the diagrams by using his conventions of relative positions for the syllogistic terms (predicate terms above their subject terms, major terms first, minor terms last, middle terms in the middle, middle terms to the left or middle or right according to syllogistic figure, etc.).

In the 17th and 18th centuries, G.W. Leibniz, Johann Lambert, and Leonhard Euler, devised and used systems of logic diagrams that were meant to do just that. They were systems for displaying the way that the conclusion of a syllogism was, in effect, contained in its premises. The terms of such inferences were taken to determine classes of individuals and were depicted as either line segments or closed curves (viz., circles). The relations between classes were modelled in the relations between the line segments or between the circles. Leibniz made use of both kinds of devices, linear and spatial, as Moktefi and Shin call them (Moktefi and Shin 2012, 629). In a paper titled "De formae logicae per linearum ductus," one of many studies unpublished during his lifetime, Leibniz wrote: "Aliquoties cogtitavi de Formae Logicae comprobatione per linearum ductus. ... propositiones per rectarum habitudines exprimentur, dum rectae rectas continent" [Several times I have thought about the comprobation of the logical form by the drawing of lines. ... propositions will be illustrated by the relations of straight lines, by the fact that straight lines contain straight lines] (Leibniz 1903, 292). The straight lines he had in mind, of course, are line *segments*. With an extensional understanding of terms, here is how Leibniz diagrammed each of the four standard categorical forms using parallel horizontal line segments for each class-determining term and vertical dotted lines to indicate the limits of the extensions of each (Leibniz 1903, 292–293):

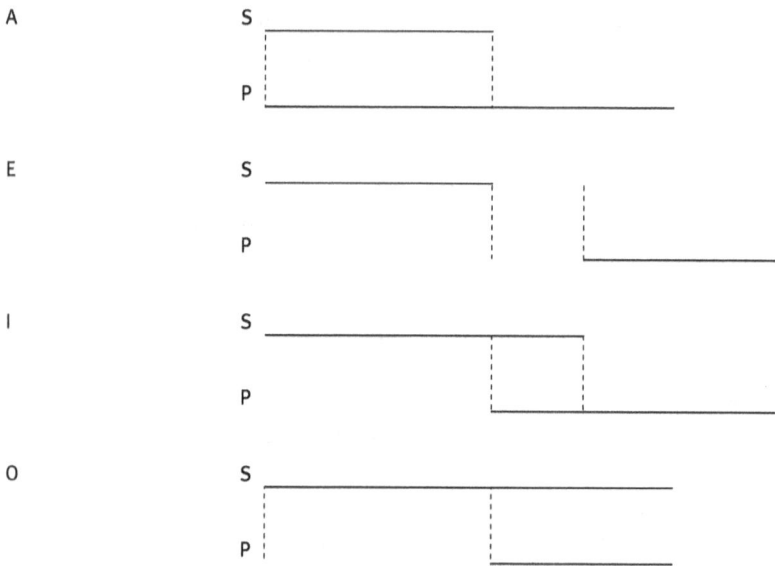

Figure 2.12: Leibniz's Linear Diagrams of the Categoricals (extensional)

Eventually, Leibniz considered an emended version of such diagrams that indicated the distribution or non-distribution of a term in a statement by doubling the portion of the appropriate line segment. For example, in an A categorical, the entire subject term line is doubled but only the portion of the predicate line lying under the subject term line is doubled; in an E categorical, both line segments are fully doubled (Leibniz 1903, 311–312). Leibniz's system also provides a visual representation of the difference between affirmative and negative statements, where the latter are seen to have at least one vertical dotted line that intersects with only one term line (Baron 1989). As well, the system exhibited the simple conversion of E and I forms by virtue of the symmetry of the respective diagrams (Belllucci, Moktefi, and Pietarinen 2014).

Of course, the true test of a system of diagrams for syllogisms is its ability to account for syllogistic inference. Can it provide a visual model of how the conclusion must follow from the premises? Here, the idea is to show that simply diagramming the premises of a syllogism is sufficient to show that the conclusion is already diagrammed. Here is an example of what a Diagram for Barbara would look like (note that the diagram is constructed by simply diagramming the two premises, while the conclusion is then simply "read off" the result):

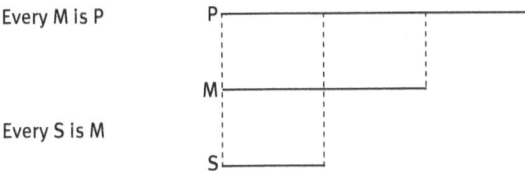

Figure 2.13: Diagram of Barbara

Leibniz demanded of an adequate system for syllogistic diagrams that the conclusion that necessarily follows be already diagrammed once the premises have been diagrammed. In this regard, he took his linear diagrams to have an advantage over his circular (Euler-like) diagrams. In particular, he held that the linear diagrams explicitly display the distributions of terms in statements. Examining this feature of Leibniz's linear system, it has been claimed that he equated distribution with quantity (Bellucci, Moktefi, and Pietarinen 2014, 24). This is a mistake unless Leibniz himself made such a blunder. Leibniz would have been well acquainted with medieval doctrines of distribution, which, though varied in details, agreed on certain basic principles. Quantity does track distribution in some cases, but not always. Terms that occur in a statement without being quantified in any way also have distribution values (i.e., are either distributed or undistributed) in that statement. In the diagrams above, Leibniz displays the distribution value of a term in a categorical statement by showing whether or not it is wholly contained in or excluded from the other term. A term is contained in another term if its entire line segment extends over all or part of the line segment of the other term; it is excluded from the other term if its line segment extends over no part of the line segment of the other term's line segment. Given any term used in a categorical statement, either it is contained in the other term, excluded from the other term, or only partially included or excluded from the other term. Such terms are distributed only in the first two cases and undistributed in the last two cases. As expected, the diagrams above show that the subject terms of A and E and the predicate term of O are distributed, while the subject terms of I and O and the predicate term of A are undistributed.

In his "General Inquires about the Analysis of Concepts and of Truths," written in 1686, Leibniz used line segments to diagram his *intensional* interpretation of statements (Leibniz 1966, 73–74). While on the extensional interpretation of a statement the relations of full/partial containment/exclusion are taken to hold between the *sets of individuals denoted* by the subject and predicate terms, on the intensional interpretation these relations are taken to hold between the *concepts connoted* by the two terms. Thus, for example, on the first interpretation,

'Every bachelor is a male' is understood as saying that the set of bachelors *is contained in* the set of males, on the second interpretation, the statement is understood as saying that the concept of being a bachelor *contains* the concept of being a male. In these diagrams the unbroken (and sometimes unnecessarily doubled or thickened) line segments indicate the range of the components of the concept determined by the term, the dashed line segments indicate *the possible* range of the components of the concept determined by the term, and the vertical line marks the limits of the ranges so indicated by line segments. Here is how the standard categoricals would be diagrammed with the intensional interpretation:

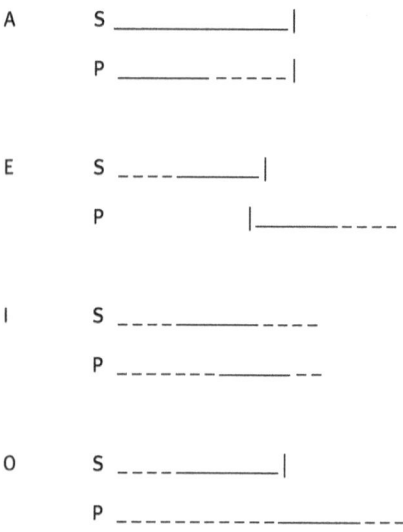

Figure 2.14: Leibniz's Linear Diagrams of the Categoricals (intensional)

These diagrams are not without elements of confusion (see, for example Parkinson 1966, xxxviii-xlii, and Moktefi and Shin 2012, 629–630).

As already noted, Leibniz devised and used not only linear but spatial (closed figure) diagrams. So, if Aristotle was not in fact the first to use linear diagrams, then a strong case could be made that Leibniz was. Unfortunately, neither Aristotle nor Leibniz is adequately credited for originating the use of logic diagrams based on line segments. That honour is usually given, if given at all, to Lambert, a polymath like Leibniz (Kant characterized Lambert as "*der unvergleichlicher Mann*"). Moreover, the consensus is that the first to make effective use of spatial diagrams was Euler. And this in spite of the fact that Leibniz had used very similar spatial diagrams well before Euler. These situations are due to a number of circumstances, not the least of which was, in spite of his exceptional prolific writing on a very wide variety of subjects, Leibniz's failure to publish more than a small fraction of his contributions. Lambert probably did not know of Leibniz's linear diagrams, but he did devise a system of linear diagrams, based on an extensional understanding of statements, that is similar to Leibniz's. Lambert (Lambert 1764) diagrammed the *known* extension of a term as a solid line segment, extending this line with dots to indicate, where needed, the term's unknown extension. As it happens, his use of such dotted lines was essential for Lambert's diagrammatic analysis of syllogisms. In diagramming a statement, he tended to place the predicate term line above the subject term line, labelling each with an appropriate letter. He also allowed a single labelled dot to indicate an individual (so that the letter label is actually to be read as a singular term). Here are examples of how Lambert would diagram the four categorical forms as well as the form of a singular statement:

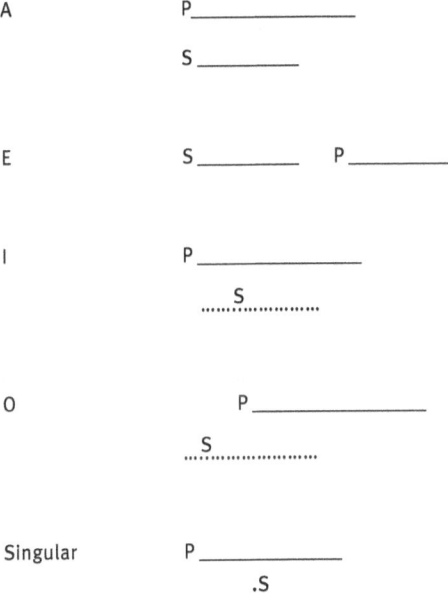

Figure 2.15: Lambert Diagrams

Recall that Leibniz's diagrams exhibited the distribution values of terms used in a statement by revealing the inclusion/exclusion of the extension of one term relative to the extension of the other. A term contained in another is shown by the line segment for the first being drawn above or below all or part of the line segment for the other term. If a term excludes another term, then the two line segments for those terms do not extend over any part of one another. Given a pair of terms used in a statement, a term that is either wholly contained in or excluded from all of the other term is distributed; otherwise it is undistributed. Lambert's diagrams can also display such distribution values. Any term whose line segment is completely solid and is wholly contained in or excluded from the other term is distributed; otherwise it is undistributed. We have also noted that Leibniz's diagrams display the conversions of E and I forms by the symmetry of their diagrammatic representations. This is not always the case with Lambert's diagrams. While the diagram for E is symmetrical the one for I is not. Nonetheless, like Leibniz diagrams, Lambert diagrams are generally adequate for exhibiting the validity of syllogisms by diagramming each premise and then reading the conclusion from the result (see Bellucci, Moktefi, and Pietarinen 2014).

As noted earlier, though Leibniz certainly preceded Lambert in the use of linear diagrams and Euler in the use of spatial diagrams, Lambert and Euler received the credit for these accomplishments well before Leibniz did. Recognition of Euler's contribution to logical diagramming was a long time coming, however. Only in the 19th century did logicians look closely at his work and come to appreciate its import and usefulness. As we will see in the next section, it was only then that significant numbers of philosophers, mathematicians, and logicians returned to the view that an adequate system of logical diagrams could be more than just a pedagogical tool for logic instruction, or a merely heuristic (but unnecessary) tool for logical tasks. Euler's circle diagrams seemed ready-made to conform to this renewed view. They are relatively easily constructed, and categorical forms are constructed from them with pairs of circles that readily exhibit, by the way the circles are arranged, the term relations that are being diagrammed. In fact, they "do not *represent* relations between classes; rather, they *have* those relations" (Moktefi 2015, 610).

In Euler's diagrammatic system, the area enclosed by a closed curve (usually a circle) represents the set of individuals in the extension of a term. A term whose extension is included in that of another term is represented as a circle inside the circle representing the second term. Pairs of terms whose extensions exclude one another are represented by circles separate from one another. Pairs of terms that share with one another a portion of their extensions are represented by partially overlapping circles. For each term, Euler used a letter label that is placed in the term's circle representation or, in the case of overlapping circles, the labels were place in the appropriate parts of those circles. Singular statements were treated as if they were universals.

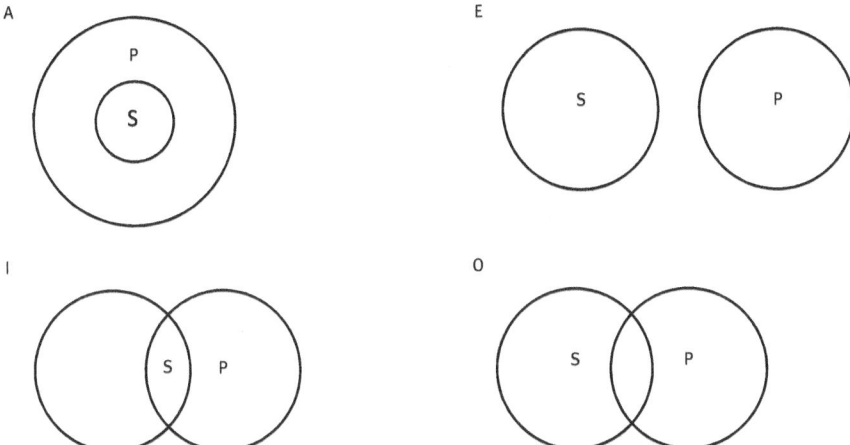

Figure 2.16: Euler Diagrams of the Categoricals

As is readily apparent, there is a certain amount of ambiguity about these representations. Do the inscribed letters indicate existence? Is an unmarked cell (delimited part of a circle) empty? While Leibniz's diagrams exhibit convertibility via symmetry, Euler's diagrams, like Lambert's, do not. Some clarity was provided by Euler's introduction of an asterisk (*) to indicate that a cell is not empty. This allowed for an alternative (and symmetric) representation of particular statements. An I categorical form could be shown as:

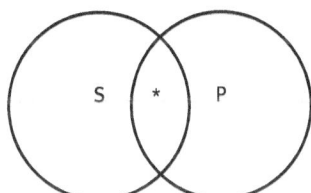

Figure 2.17: Euler's Asterisk for 'Some S is P'

In spite of their limitations, Euler's circles can be used to shed useful light on syllogistic reasoning. Again, the principle underlying the system was that diagramming the premises of a syllogism was sufficient to reveal that of the conclusion was thereby represented. Here is an example of a Euler diagram for a Cesare syllogism ('Every S is M; no P is M; therefore, no S is P'):

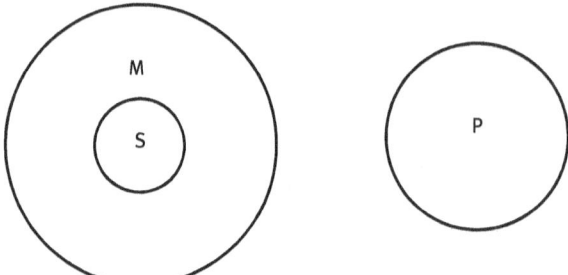

Figure 2.18: Euler Diagram for Cesare

Euler's system of logical diagrams became the starting point for Venn's development of his diagrammatic system. Indeed, Euler diagrams became the standard graphical tool for analyzing syllogistic reasoning for many decades and remains today an inspiration for, and a subject of, intense study, particularly in relation to subsequent diagrammatic systems such as Venn's (see, for example, Hammer and Shin 1998, Savio 1998, Stapleton and Masthoff 2007, Sato, Mineshima and Takemura 2011).

2.3 Venn, Peirce (and Frege?)

> He makes diagrams of spots connected by lines; and it is easy to prove that every possible system of relationship can be represented, although he does not perceive the evidence of this.
>
> Peirce
>
> [Venn's system of diagrams] affords no means of exhibiting reasoning, the gist of which is of a relational or abstractional kind. It does not extend to the logic of relations
>
> Peirce
>
> A diagram ought to be as iconic as possible, that is, it should represent relations by visible relations analogous to them.
>
> Peirce
>
> Speech often only indicates by essential marks or by imagery what a concept-script should spell out in full. At a more external level, the latter is distinguished from verbal language in being laid out for the eye rather than for the ear.
>
> Frege
>
> I believe I can make the relationship of my *Begriffsschrift* to ordinary language clearest if I compare it to that of the microscope to the eye.
>
> Frege

John Venn introduced his version of spatial diagrams for logic by noting how familiar Euler's circles had become even to those untrained in logic (Venn 1880, 1). In the field of logical diagramming, while the century that began with Euler was dominated by his circles, the century that began with Venn was dominated by *his* interlocking circles. Even in the 21st century, school children have learned something of Venn diagrams. Today, if one thinks about logic diagrams at all, one probably has in mind Venn diagrams.

Venn's aim in devising his system of logical diagrams was to provide a visual, graphic mapping of Boole's new algebraic logic, a logic adequate for the analysis both of inferences involving sentences made up of monadic terms and of inferences involving unanalyzed sentences – a logic of terms and a logic of sentences. Euler's diagrammatic system was inadequate for the latter task. Venn, as well as many logicians who followed, discerned a number of other inadequacies and limits that seemed to plague Euler's system. A common critique of that system was that Euler was unable to represent every possible relation between two terms. Sun-Joo Shin (Shin 1994b) has given a clear, concise account of the most important limitations on Euler's system.

Her first example shows that there is a degree of ambiguity in Euler diagrams. Consider these diagrams:

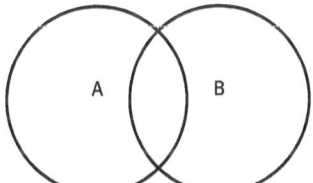 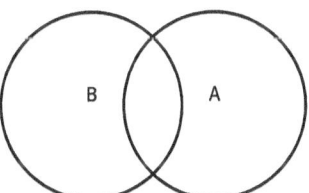

Figure 2.19: Euler Diagrams for 'Some A is not B' and 'Some B is not A'

Given Euler's convention that the placement of a term letter in a cell represents the location of part of the extension of that term, each diagram could be taken to show both that some As are not Bs and some Bs are not As. This is an ambiguity that "is a fatal flaw in a representation system" (Shin 1994a, 14). Shin's second criticism is that Euler diagrams fail to provide a perspicuous representation of contradictoriness, such as that between A and O categoricals. For example, 'All A is B' and 'Some A is not B' are contradictory. Yet these Euler diagrams for them do not show this in an "obvious way" (Shin 1994b, 15):

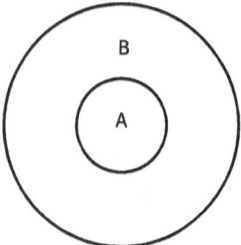 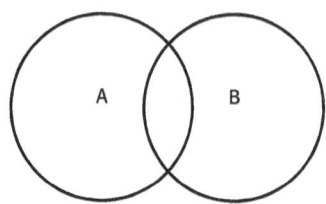

Figure 2.20: Euler Diagrams for Contradictory Pairs

Finally, Shin offers two examples of how Euler diagrams perform poorly when it comes to the task of combining more than one piece of information in a single diagram (Shin 1994b, 15–16). Consider the Disamis syllogism 'All A are B; some A are C; so some B are C'. Euler diagrams for each premise can be constructed like this:

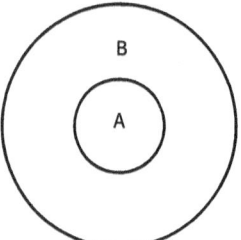 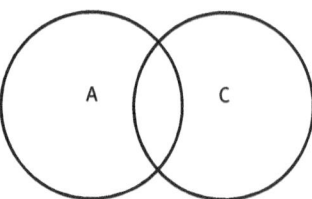

Figure 2.21: Euler Diagrams for 'All A are B' and 'Some A are C'

There seems to be no way the information conveyed in these two diagrams can be combined in a single diagram that would represent the conclusion. The same limitation applies to attempts to show by Euler diagrams that an A and an E categorical are logically contrary but not contradictory. They can be diagrammed as follows:

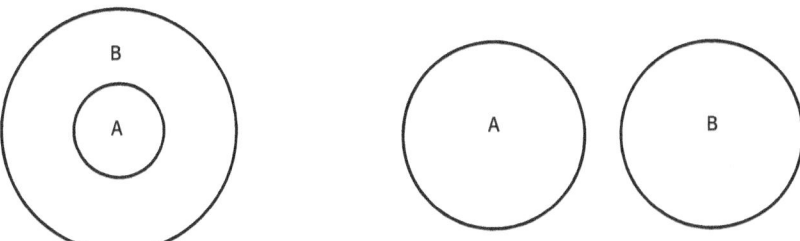

Figure 2.22: Euler Diagrams for Contrary Pairs

The diagrams cannot be combined and, indeed, appear to show incompatibility between the two statements. However, when the A term is empty (when there are no As), they are not only compatible but, in effect, identical.

The inability of Euler diagrams to provide a perspicuous way to combine diagrams has been a persistent criticism, "the most serious problem of this system" (Shin 1994b, 15). It is the main reason for the ascendency of Venn's diagram system over Euler's. Venn diagrams are built to be combined. This is due to the fact that Venn diagrams allow for the representation of incomplete information. It is possible to add new information to an existing diagram, as when a second premise is added to an original premise. While Euler devised his system of diagrams as a tool for exhibiting syllogisms, Venn, because he sought to provide a visual analogue for Boole's algebraic logic, devised his diagrams as a tool adequate for exhibiting all kinds of deductions, including syllogisms.

A "primary" Venn diagram is a pair of interlocking closed plane figures (usually circles or ellipses). Typically, a Venn diagram for a syllogism begins by inscribing inside a larger rectangle (representing the universe of discourse) three primary circles representing the extensions of the three terms of the syllogism. The border of each circle intersects exactly twice with each other circle, resulting in eight distinct cells (including the area outside the circles but within the universe of discourse). Each circle is given a term letter label. One premise is diagrammed by a primary diagram representing the relation between the extensions of its terms (the middle term and an extreme term). A third circle, representing the relation between the extensions of its terms (the other extreme and the middle) is added to this primary diagram. Given the validity of the argument, the conclusion, indicated by the relations between the extensions of its terms (the major and minor terms), can then be seen to have been already diagrammed. A set (circle) is shown to be empty (have no individual in its extension) by being shaded. It is shown to be non-empty (have at least one individual in its extension) by having an 'x' placed in it. Here are two examples of Venn diagrams for syllogisms:

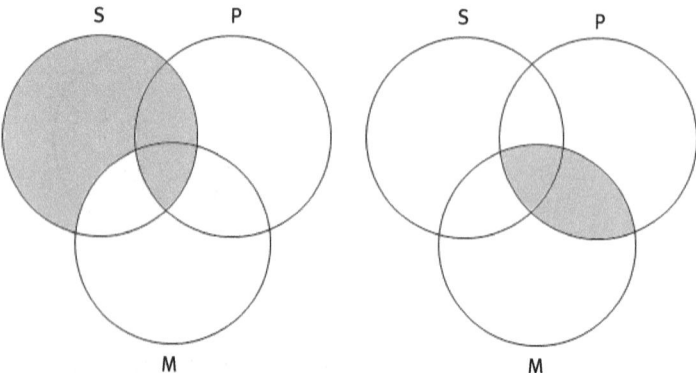

Figure 2.23: Venn Diagrams of Celarent and Festino

As is happens, though it is now standard to do so, Venn did not present his diagrams by inscribing them inside a rectangle representing the universe of discourse. He simply took the (undemarcated) area outside his circles to be understood as the universe of discourse. Lewis Carroll, in his inimitable style, responded to this:

> My Method of Diagrams *resembles* Mr. Venn's, in having separate Compartments Assigned to the various Classes, and in marking these Compartments as *occupied* or as *empty*; but it *differs* from his Method, in assigning a *closed* area to the *Universe of Discourse*, so that the Class which, under Mr. Venn's liberal sway, has been ranging at will through Infinite Space, is suddenly dismayed to find itself "cabin'd, cribb'd, confined', in a limited Cell like any other Class! (Carroll 1958, 176)

Also, and again now standard in the construction of Venn diagrams, Venn did not use an 'x' to indicate existence. That was one of Peirce's innovations. Nonetheless, it has been shown that a system of such Venn diagrams can serve as a decidable, sound, and complete formal logic of monadic terms (Hammer 1995, 79–81).

Boole's algebraic logic allowed for multiple interpretations. In particular, the variables of the symbolic language could be interpreted not only as terms (whose extensions are sets) but as entire statements (propositions). It was a propositional logic as well as a (monadic) predicate logic. Venn's diagrammatic logic was aimed at mirroring Boole's algebraic logic, and in this respect it did. A Venn closed curve could be interpreted as representing either the possible extension of a term or the possible truth-value of a proposition (Gardner 1982, 48–54). So Venn's system of logic diagrams had significant advantages over Euler's. Yet it also faced some important limitations.

The limitation that first drew attention from logicians was the difficulty of using closed figures to represent more than a relatively small number of terms. One can use three interlocking Venn circles for the three terms of a syllogism; one can use four appropriately arranged ellipses. But beyond that, accurate diagrams depicting more than four terms soon become more complex, less perspicuous, and eventually require a given set to be represented by discontinuous closed curves. A number of logicians (especially Allan Marquand and then Carroll) made some progress by substituting rectilinear shapes for Venn's circles. However, these kinds of solutions tend to result in the loss of symmetry and simplicity, not to mention perspicacity, to the extent that the initial *raison d'être* of logical diagramming recedes into the distance (see, for example, Moktefi, Bellucci and Pietarinen 2014). Another limitation of Venn's diagrams was due to its failure to explicitly label negative terms. The extensions of a term and its negation (complement) determine a set and its counter-set. Euler and Venn systems simply adopt the convention that any point outside of a labelled figure represents the counter-set of the set represented by that figure. Both Marquand (Marquand 1881) and Carroll (Carroll 1886; see also Abeles 2007 and Moktefi 2008 and 2013) were able to introduce larger numbers of terms into their diagrams by simply dividing the square or rectangle representing the universe of discourse into two halves representing a term and its negation, with such division continuing orthogonally to the preceding division for each additional term. This practice eventually sacrifices continuity, but it allows for the representation of any finite number of terms. Here is an example of such a "Square" diagram:

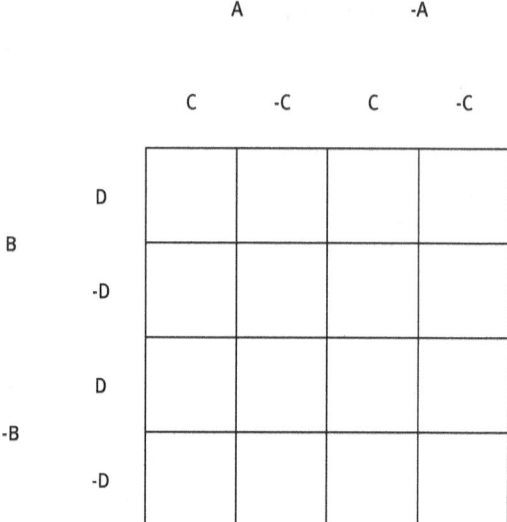

Figure 2.24: Square Diagram for 4 Terms

Genuine progress came only with Peirce's development of his "Existential Graphs" (Peirce 1933). As he saw it, Venn's system may have been plagued by difficulties in providing diagrams involving more than a handful of terms, but it also had more fundamental expressive limitations. Among these are its inability to graphically depict existential statements, disjunctions and relational statements (Shin 1994b, 21). Peirce's solution to the first of these was to introduce the inscription of a letter (usually 'x') in any cell taken to be nonempty. As for the second of these limitations, there are two conditions to consider but a single solution. Sometimes one wants to claim that a cell with sub-cells is nonempty. Consider a Festino syllogism. The minor premise requires that the cell representing the intersection of the extremes be nonempty. But that cell consists of two sub-cells. An 'x' must be placed in at least one of these, but we don't know which one. Peirce's solution is to put an 'x' in each and then connect them by a straight line segment (see Moktefi and Pietarinen 2015). Thus the diagram shows that our information here is incomplete, we only know that one sub-cell *or* the other is nonempty. Peirce also replaced Venn's shading with the inscription of 'o' to represent emptiness.

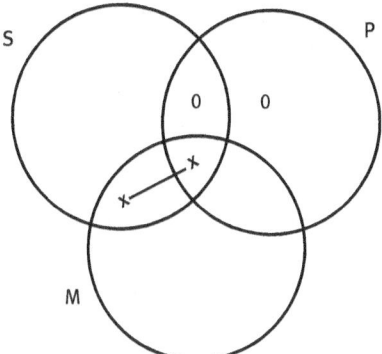

Figure 2.25: Peirce's Diagram for Festino

This same device of using a straight line segment to connect alternatives can be used as well to diagram disjunctive statements. In such cases, a disjunction can be graphed by simply connecting the two separate graphs representing each by using his straight line segment (Shin 1994b, 116 ff).

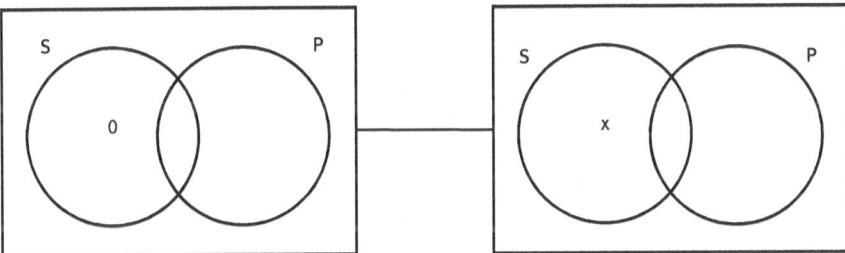

Figure 2.26: Pierce Diagram of Propositional Disjunction

It is fair to say that the surge of interest in logic diagrams by philosophers in recent years has been propelled in a significant way by the re-discovery and subsequent study and exploitation of Peirce's work on Existential Graphs (see especially Roberts 1973 and Zeman 1964, but also Shin 1994a, 1994b, and 2002 and Hammer 1995). Peirce went far beyond simply making amendments such as those mentioned above to Venn's diagrams. While Venn had intended his system as a visual analogue of Boole's algebraic logic, Peirce went much farther. Boole's symbolic system allowed for interpretation as a logic of unanalyzed statements (a propositional logic) and for interpretation as a logic of monadic predicates (essentially, syllogistic). By contrast, Peirce envisaged a system of logical diagrams that would *be* a formal logic in its own right, not just an analogue or reflection of any symbolic, non-visual system of logic. He thought of his graphic logic as his most important contribution to philosophy. Yet he is primarily known to most logicians today as one of the two founders of modern predicate logic (MPL), the other, of course, being Frege. Peirce's *symbolic* logic included a logic of unanalyzed proposition and, resting on that logic, a logic of predicates (both monadic and relational) with explicit quantifiers as well as identity. It is now known to be both sound and complete.

Peirce's *graphic* logic was intended to be *iconic*, where the relations expressed between propositions and the relations expressed between predicates are depicted visually by figurative elements whose relations are analogous to those logical relations. The distinction between symbols and icons was part of Peirce's overall theory of *signs*. While an icon denotes or represents whatever it represents by resembling it, a symbol's representation is fixed by a convention (an agreement among its users to give it that representation). As it happens, Peirce's logic graphs, though primarily iconic, also involve symbols (for propositions and for predicates). So it is a *heterogeneous* system of logic. Indeed, Peirce was "the first person to establish the theoretical grounds for heterogeneous systems" (Shin 2002, 34).

What follows is merely a brief sketch of some of the main features of Peirce's iconic logic (fuller accounts are found in the works cited above). Peirce saw his graphic logic as consisting of various parts, the first two of which ("Alpha" and "Beta") amount to his propositional logic and his predicate logic, respectively. In each case, a graph is always inscribed on a "sheet of assertion" (a part or whole of a blank page) that is, in effect, meant to represent the appropriate universe of discourse. The vocabulary of Alpha consists of an unlimited number of sentential/propositional symbols (A, B, ...), an icon indicating negation (a "cut" that is simply a closed figure, and the juxtaposition of pairs of graphs (indicating conjunction). It is well known that negation and conjunction are sufficient for formulating a logic of propositions (indeed, well before Sheffer, Peirce showed that a single propositional connective would be sufficient). Here are some examples of Alpha graphs:

A and B: A B

Not A:

Not both A and B:

If A then B:

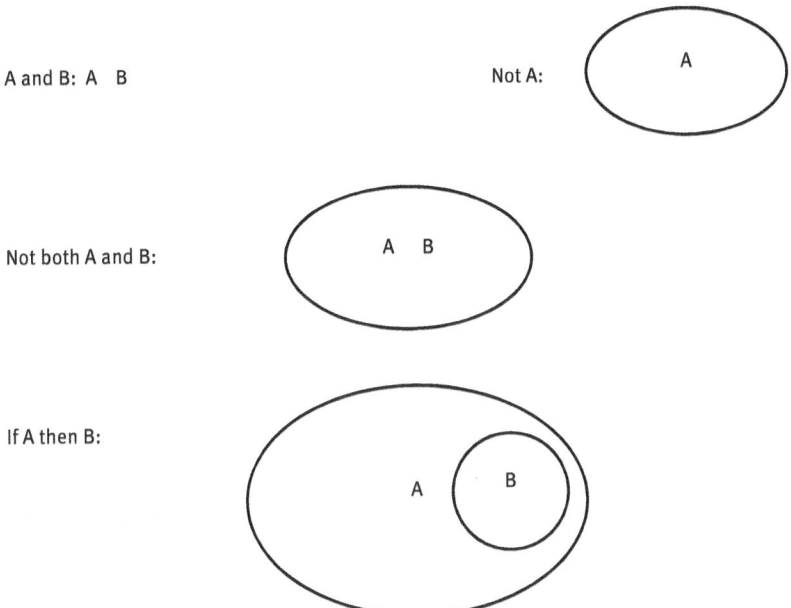

Figure 2.27: Alpha Graphs

In the vocabulary for Beta, the letter symbols represent predicates, the cut represents negation, and "lines of identity" (a straight or curved line segments that can branch in various ways). In this system cuts are allowed to cross lines. A line of identity can be read as 'something is'. Thus a line segment that has a predicate symbol at one of its endpoints represents the assertion that something has the property represented by that predicate. A line having a predicate symbol at two endpoints represents the assertion that something that has the first property is something that has the second. Here are Beta graphs for the four standard categoricals:

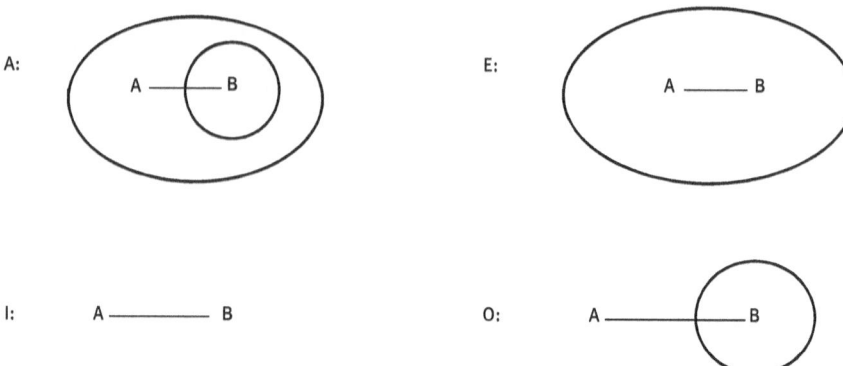

Figure 2.28: Beta Graphs for the Four Categoricals

No system of logic diagrams before Peirce could offer a way of adequately diagramming relational sentences. Relationals had always been a serious challenge to traditional logicians so that diagrammatic systems were generally confined to analyzing categorical statements and syllogistic arguments. But Peirce was not interested in simply providing a graphic version of Boole's logic (a logic that avoided relationals). He was intimately familiar with De Morgan's attempts to build a logic of relations. In fact, the main title of one of Peirce's earliest and most important logical studies was "Description of a Notation for the Logic of Relatives" (Peirce 1870; for more on Peirce's introduction of relationals into his system of logic see Brady 1997, Burch 1997, and Merrill 1997). Eventually Peirce had to incorporate such expressions into his Beta graphs. But, as Shin has written, "It is well known that relations are more difficult to represent in a graphical system than properties" (Shin 2002, 53). Peirce made a number of attempts, but his final solution takes relational predicate symbols as occupying multiple line endpoints. Here are some examples of graphs involving relational predicates:

Some boy admires a footballer.

A man gave a rose to a woman.

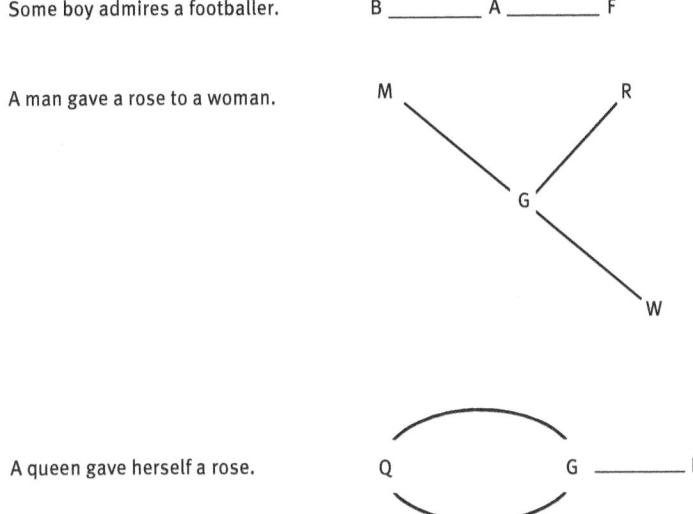

A queen gave herself a rose.

Figure 2.29: Beta Graphs with Relationals

Identity/nonidentity is easily graphed. The graphic representation of 'Some A is identical to itself' is represented by placing A at both endpoints of a curved line; 'Some A is not identical to itself by placing A at one endpoint of a line segment and placing a cut around the other endpoint; and 'Every A is identical to itself' is represented by placing the entire graph for the previous statement within a cut. Thus:

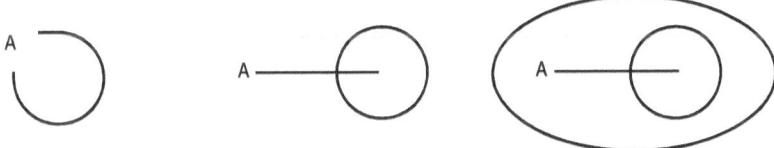

Figure 2.30: Beta Graphs of Identity

For any formal system that has the expressive power to formulate sentences with relationals, which necessarily opens the possibility of multiply quantified forms, it is natural to ask how the possible entanglement of multiple quantifiers is managed. Beta graphs solve the problem in a fairly straightforward way. First, recall that lines of identity *simpliciter* amount to existential quantifiers, so the guiding principle governing quantifiers is that a line whose outermost part is enclosed by an even number (including zero) of cuts is read existentially; otherwise it is read universally. These two graphs illustrate the difference:

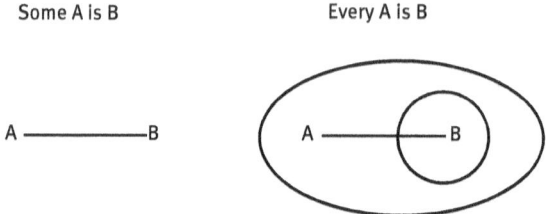

Some A is B   Every A is B

Figure 2.31: Beta Graphs for Quantification

Next, when multiple quantifiers are involved, quantifier scope is determined by the extent of its enclosure: the less enclosed a line is, the larger its scope. These two graphs illustrate this:

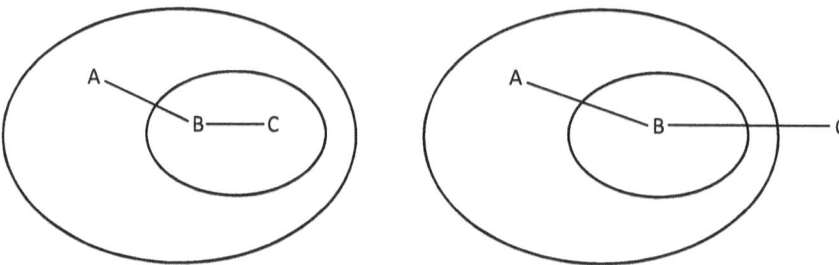

Figure 2.32: Beta Graphs Illustrating Quantifier Scope

Note here that in the first graph, A, unlike B and C, is enclosed only once (so it has largest scope) and it is oddly enclosed (so it is universally quantified). In the second graph, C is enclosed zero times (so it has largest scope) and it is evenly enclosed (so it is existentially quantified).

It has been noted often that Peirce's graphic logic does not involve icons (or even symbols) for individuals. His procedure was to take such a term (e.g. a name like 'Obama') as elliptical for thing identical to the thing named (thus: 'thing that is (identical to) Obama'). This is something like Quine's pegasizing procedure for turning names into predicates. But there are limitations to such a practice and the solution is to admit symbols denoting individuals into Beta graphs. B.H. Slater has outlined the limitations involved and this solution (Slater 1998).

Peirce's Existential Graphs have both the expressive power and deductive power of modern first-order predicate logic with identity. As with other graphic systems of logic, such as Venn's, inference is shown by exhibiting how a conclusion is already diagrammed once the premises have been diagrammed. But one of the most important features of Peirce's icon-based diagrammatic logic is that his graphs allow "multiple readings" (Shin 2002, 75ff). For example, even though the vocabulary and syntax for Alpha graphs admits direct representation for only negation and conjunction, conditionals and disjunctions can readily be represented without recourse to further notational or graphic devices. By contrast, symbolic formulae permit just one reading.

> Hence, for a given diagram we may obtain more than one sentence as its translation. ... The main reason why multiplicity [of readings] has been ignored is, we suspect, that diagrams have been considered as a form of symbols, and hence, no attention was paid to fundamental differences between diagrammatic and symbolic representation. ...Diagrams, being spatial, may be perceived in more than one way, as the classic Gestalt phenomenon illustrates. (Moktefi and Shin 2012, 665–666)

This makes graphic/diagrammatic systems not only more perspicuous but also more efficient. For example, it is possible to *show* that a number of logical relations among a set of terms or propositions are equivalent with a single diagram rather than *proving* each equivalence with a series of symbolic proofs. Here is a very simple example. These two formulae are logically equivalent: ∃x∃y(Ax & Bx & Rxy) ['Some A is R to some B'] and ∃y∃x(By & Ax & Rxy) ['Some B is R'ed by some A']. Any proof of this requires the application of rules for exchanging quantifiers and rules of conjunction commutation. A Beta graph permits both readings at once, without further ado.

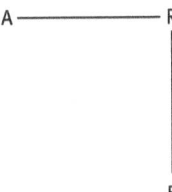

Figure 2.33: A Single Beta Graph Illustrating a Logical Equivalence

Russell famously claimed that a good notation is like a live teacher; that surely holds for a good diagram as well.

Comparisons of the contributions to symbolic logic by Peirce and by Frege are inevitable and valuable. We know now that Peirce developed an effective system of logical diagrams (though recognition of this took a relatively long time). We know also that he developed key elements of what is now the standard first-order predicate calculus, introducing the symbols for universal and existential quantifiers. He shares honours for this second accomplishment with Frege. But what of Frege's symbolic language for logic? Frege's contemporary critics, in particular Schröder (Schröder 1880), took his *Begriffsschrift* notation to be simply too confusing, too difficult to print, too space consuming, too *outre*. Subsequently, logicians such as Russell saw Frege's notation as nothing more than an awkward variation of Peano's notation, a version of which became standard for mathematical logic (Macbeth 2013, 43). But recently Danielle Macbeth (Macbeth 2012a, 2012b, 213) has shown that in fact Frege's aim all along was not to devise a system of symbolic logic but instead a diagrammatic system for mathematical reasoning. So while most logicians regard Frege's concept-script as merely a two-dimensional variant of a one-dimensional (and more convenient) *symbolic* logical notation, others see it as a genuine *diagrammatic* system for *doing* (not simply *recording*) deductive reasoning – at least in mathematics. "Because the notation functions graphically, Frege can exhibit the contents of mathematical concepts in a way enabling deductive reasoning on the basis of those content." (Macbeth 2012b, 80). And his notation goes beyond just exhibiting conceptual content. "Rather one reasons in the language; the language is a vehicle or medium of reasoning." (Macbeth 2012b, 63).

> In Frege's case ... we see (perceive) the diagram, and therefore are inclined to see further (intuit) through the diagram, into the objective domain of concepts (which is of course not to say that this is how we primarily get access to this domain). (Toader, 2004, 24).

Noting that Boole had required signs for addition, subtraction, multiplication and equality (as well as 0 and 1), Frege said that his notation consisted only of:
1. The horizontal 'content-stroke',
2. The negation-stroke,
3. The vertical stroke that combines two content-strokes, 'the conditional-stroke',
4. The vertical judgement-stroke. (Frege 1979, 52)

Frege called attention to two, among a number of other, difference between his new language and Boole's formal language. "The first thing to notice is that Boole uses a greater number of signs." (Frege 1979, 35). Moreover, while Boole made use of familiar mathematical symbols for his language, Frege eschewed such symbols. "I follow the basic principle of introducing as few primitives as possible, not from any aversion to new signs – but because it makes it difficult to survey the state of a science if the same thing is dressed up in different garbs (Frege 1979, 36). The logical principles that govern reasoning in mathematics cannot be stated in a language that makes use of the primitive symbols of mathematics itself. And, of course, another difference between Boole's logic and Frege's is that Boole and his followers saw a logic of terms (viz, traditional syllogistic) and a logic of propositions as distinct; by contrast, Frege "avoid[ed] such a division into two parts" (Frege 1979, 14), reducing the former to the latter. This reduction is the exact opposite of Leibniz's as we have seen. Frege's reduction here is a consequence of what is now known as *Frege's Priority (or Context) Principle*, which we will discuss more fully in Chapter III. Finally, we saw that Frege constructed his system for the representation, articulation, and the practice of reasoning concerning relations that hold among mathematical *concepts* (thus a "concept-script" or *Begriffsschrift*). From his point of view, "The Boolean signs (in part stemming from Leibniz) are completely unsuited to this, which is scarcely to be wondered at when you consider their purpose; they are merely meant to present the logical form with no regard whatever for the content" (Frege 1979, 47).

Frege's two-dimensional concept-script was linear. Here are some simple examples of Fregean diagrams, first for statements in *propositional* logic (which he and his followers take to be primary logic, foundational for *predicate* logic) followed by examples of diagrams for statements in predicate logic. In each case, diagrams will be accompanied by their more familiar counterparts in the symbolic language of standard mathematical logic.

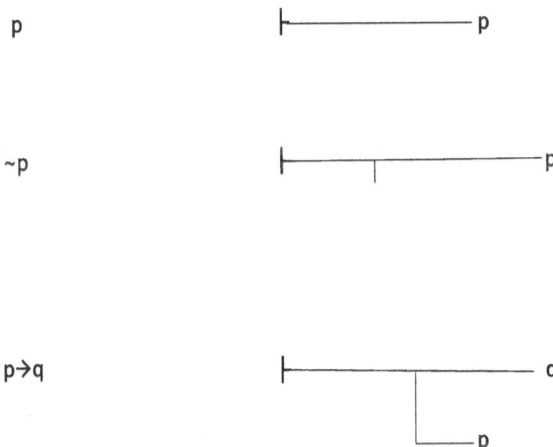

Figure 2.34: Fregean Diagrams for Propositional Logic

Let Φ(x) and Ψ(x) be formulae using x:

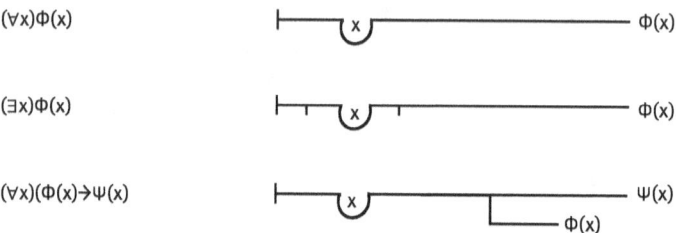

(∀x)Φ(x)

(∃x)Φ(x)

(∀x)(Φ(x)→Ψ(x))

Figure 2.35: Fregean Diagrams for Predicate Logic

The notation was perspicuous in the sense that a single diagram could often represent a number of logical relations. For example (from Moktefi and Shin 2012, 660–661) the logical relations (C & B) →A, C→(B→A), (B&C)→A, and B→(C→A) can be seen in the following:

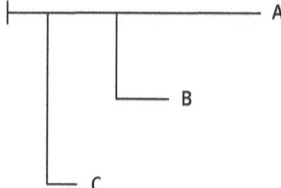

Figure 2.36: Fregean Diagram

One must always keep in mind that Frege's intention was to build a formal language that would be free of the flaws of natural language and adequate as a medium in which to conduct reasoning in mathematics (in particular, arithmetic). Moreover, mathematics is a science and "[t]he goal of scientific endeavor is *truth*" (Frege 1979, 2). Indeed, "the laws of logic are nothing other than an unfolding of the content of the word 'true'" (Frege 1979, 3). So, a logical deduction or proof must establish a truth from other, given, truths. Ultimately the truths of a science are derived from a relatively small set of *primitive* truths (axioms) that are themselves in no need of proof, unprovable. The axioms of mathematics are the *laws of logic* and "the whole of mathematics is contained in these primitive truths as in a kernel" (Frege 1979, 205). The entire account of the nature of judgments and their content (propositions, thoughts) and the concepts that can be gleaned from such propositions, as well as the new language for expressing these, merely set the stage for the process of logical inference of truths from given truths – proof. "The theory of concepts and of judgement is only preparatory to the theory of inference" (Frege 1979, 175). For Frege, proof in mathematics is a special case of logical proof. "*Mathematics has closer ties with logic than does any other discipline*; for almost the entire activity of the mathematician consists in drawing inferences" (Frege 1979, 203).

While there is no doubt that Frege's concept-script offers a graphic (rather than merely symbolic) representation of propositions and concepts, it is his method of exhibiting logical inference that eventually raises questions about whether this really amounts to a genuine system of logical diagramming. His proof theory takes all logical inference (and thus all mathematical inference) to proceed from given judgments to new judgments. The theory invokes his nine axioms (which he allows can be readily converted to rules) and a small number of explicit rules of inference for both propositional and predicate logic. The primary rule is simply modus ponens, but the system also makes use of other rules, including those that amount to such now-familiar rules as universal generalization and instantiation, and substitution (of equivalents). Here are two examples:

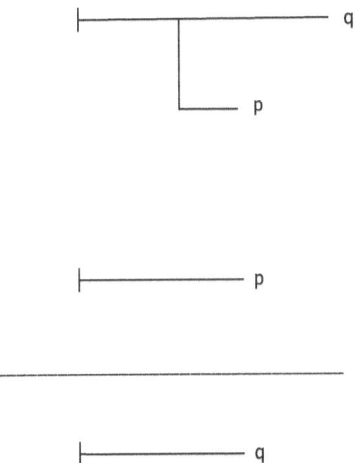

Figure 2.37: Fregean Modus Ponens

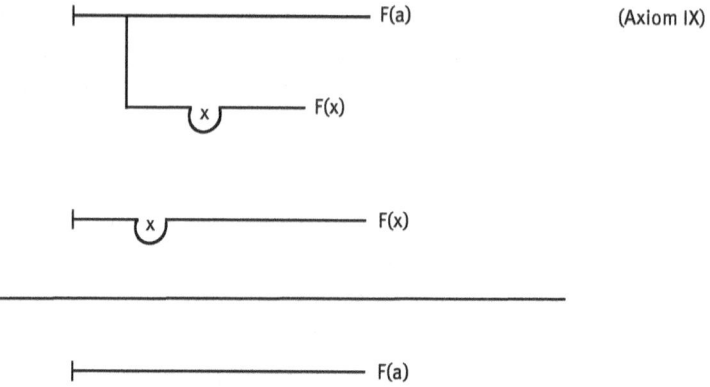

Figure 2.38: Fregean Universal Instantiation

(Note that this is, in effect, and example of modus ponens.) The long horizontal line separating conclusions from premises was used by Frege without explanation (but see Kanterian 2012, 75 ff). But the obvious question here concerns how such graphic representation of premises and conclusions support anything like *visual* inference. How is the representation of the conclusion already *seen in* the representation of the premises? Consider this valid argument form:

1. Every A is B.
2. Some A is not C.
3. Every B is D.

So, 4. Some D is not C.

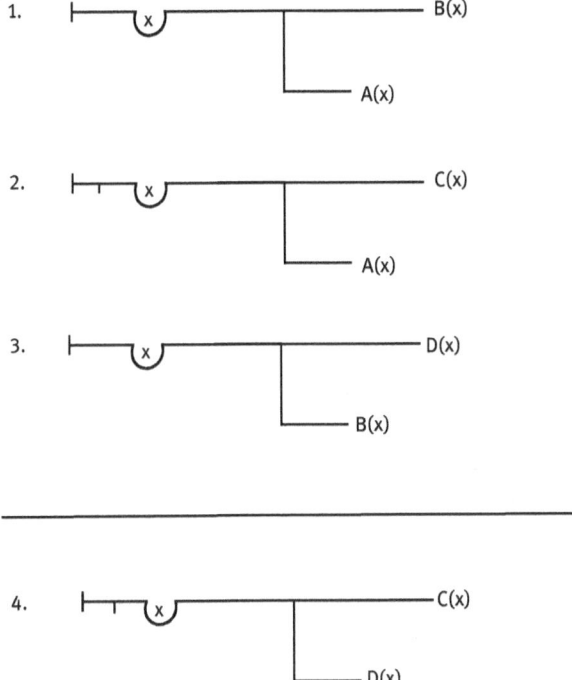

Figure 2.39: A Fregean Proof

Contemporary mathematical logicians are heirs to the legacy established by Frege's revolution in logic. Though a number of different notational systems are now in use, few logicians today have close familiarity with Frege's two-dimensional notation, and fewer still make any use of it. However, their methods of formal deduction are essentially Frege's. In such proof no attention is paid to the possibility of visually exhibiting inferential steps. And that was also true for Frege's inferences. Nevertheless, a new generation of logicians, philosophers, cognitive psychologists, and information theorists have come to recognize what logicians from Leibniz to Venn to Peirce saw: that symbolic representation is not the only viable kind of representation, and that inferences carried out in a symbolic medium can often be carried out (even more perspicuously) in a visual system as well (but regarding Peirce, see Lemon 1997, 215).

2.4 "The Renaissance of Diagrammatology"

> With the aid of visual imagination we can illuminate the manifold facts and problems of geometry, and beyond this, it is possible in many cases to depict the geometric outline of the methods of investigation and proof without necessarily entering into the details connected with the strict definitions of concepts and with the actual calculations.
>
> Hilbert

> Diagrams, like sentences, carry information: they carve up the same space of possibilities, though perhaps in very different ways. A good diagram, for example, may represent information in a form that is particularly appropriate for the subject matter at hand, one that allows you to visualize and manipulate the information more readily than would a collection of sentences or even a different sort of diagram.
>
> Barwise and Etchemendy

In assessing recent work on the nature and use of diagrammatic techniques for the analysis of logical deduction, Sun-Joo Shin has been called "a leading figure of a recent *Renaissance* of diagrammatology (the study of applying diagrams for representing logical relations)" (Danka 2016, 151). And *Renaissance* it surely is. The contrast between the volume, breadth, innovation, and quality of research concerning all aspects of diagrammatic reasoning prevalent since the 1980s is in stark contrast to the virtual absence of (often hostility to) such research during the eight decades or so that preceded it. While Peirce saw the great value of using graphic techniques for carrying out logical tasks (especially accounting for inferences involving relationals) and even considered his system of Existential Graphs more fundamental and efficient than even his own symbolic techniques, Frege had a much more limited view about the value of his own visual system. We know why those two diagrammatic systems failed to influence mainstream logic for so long. Frege was offering something far too space consuming, complex, confusing, cumbersome; Peirce was just too far away. But why did no other philosopher, logician, or mathematician at least consider the possibility of examining the logical potential of a system of diagrammatic reasoning? What delayed the Renaissance?

As Amirouche Moktefi and Sun-Joo Shin have pointed out so clearly, such unexpected challenges to mathematics and logic as the development of non-Euclidean geometries in the late 19th century, the paradox that Russell found at the heart of Frege's logic (not impacted by his notational system) at the beginning of the 20th century, and the discoveries of incompleteness for consistent formal systems by Gödel later on, "pushed the entire discipline to be more obsessed with fomalization than ever before. And the more formalization has been demanded, the more prevalent has been symbolization" (Moktefi and Shin 2012, 672). Logicians came to understand that the only defense they saw against the mathematical and logical challenges mentioned above was formal accuracy, (i.e., rigor), and rigor can only be guaranteed via symbolization. Diagrammatic (so non-symbolic) reasoning is not rigorous – so not formalizable. However, both Peirce and Frege had provided evidence in their own diagrammatic systems that it is possible to achieve rigor, in terms of accuracy and efficiency, not to mention perspicacity, in a system that is decidedly non-symbolic. But all this was ignored in the rush to establish the appropriate defense against the challenges to the foundations of mathematics and the paradoxes of logic. A defense was seen as available only when armed with a system of logic that was formalizable, thus symbolic.

It took some time, but eventually the "renaissance of diagrammatology" began. Near the end of the 1980s, Jon Barwise and John Etchemendy (see Barwise and Etchemendy 1991, 1993, 1995, 2002) initiated the *Heterogeneous Logic* project. The now well-entrenched logic of the academy treated all reasoning as carried out in the mode of sentences (symbol). Such logic is *homological*, admitting only a single mode of reasoning. Yet, even a cursory survey of the ways ordinary people carry out reasoning tasks shows that they make use of a variety of media. Sentences, of course, but also pictures, maps, charts, analog clocks, family trees, flow charts, models, diagrams, etc. The logic that people in ordinary situations use is most often *heterogeneous*. The recognition of heterogeneous reasoning entails the recognition of the possibility of formulating a viable system of logical diagramming. "One can have rigorous logically sound (and complete) formal systems based on diagrams" (Barwise and Etchemendy 1995, 214).

As Danka said, Sun-Joo Shin is a leading figure here. Her book *The Logical Status of Diagrams* (Shin 1994b) was a further milestone in the development of the project of heterogeneous logic. There, she provided a valuable case study of a heterogeneous system of logic that is both expressively and deductively powerful and both sound and complete. Diagramming for logical purposes must rely on something like perceptual inference. However, there is an assumption that a visual system is expressively weaker than a linguistic/symbolic system. Shin argued (Shin 1994b, 153 ff) that representational systems in general are characterized by a degree of convention as well as a degree of reliance on perceptual inference; and conventionality is inversely proportional to perceptual inference. Diagrams may have a lower degree of resemblance to their target objects than, say, photographs, but they have a much higher degree of resemblance than linguistic/symbolic representation, which has no resemblance to its target object. Compare "Every member of a set is a member of any set the former set is included in" with a simple Euler diagram. Diagrammatic representation requires few conventions, allows for immediate understanding, relies on perceptual inference, and is more natural than purely linguistic or symbolic representations. Shin's accomplishment was considerable and a valuable contribution to the establishment of the value of diagrammatology. Nevertheless, there was more to do. Her case study offered a system of logic that was exclusively monadic, eschewing the incorporation of relationals (see Englebretsen 1998). Fortunately, she has continued to explain, defend, and expand heterogeneous logic. In 2002, she demonstrated the power of Peirce's Existential Graphs when understood as a system of heterogeneous logic, showing that he was "the first person to establish the theoretical grounds for heterogeneous systems" (Shin 2002, 34). She continued her work on heterogeneous logic in her 2004 "Heterogeneous Reasoning and its Logic" (Shin 2004). In effect, a heterogeneous system of logic is a visual, diagrammatic system that incorporates elements from non-diagrammatic media (especially symbolic). Such systems usually augment diagrams with labels. For example, a typical Venn diagram will use a letter label for each enclosed curve. Such heterogeneous logics have a better chance of more closely modelling how reasoning is ordinarily carried out than homological logics. The "heterogeneous logic project" has continued to grow. Logicians have been joined by mathematicians, computer scientists, cognitive psychologists, and others in order to study and address a wide variety of aspects of heterogeneous reasoning (see Moktefi and Shin 2012, 676–677).

We have travelled in time (perhaps too rapidly) from Aristotle's day to our own, stopping briefly now and then on the way to catch a glimpse of the results some logicians have obtained in the effort to develop workable systems of logical diagrams. One thing to notice is that almost all of those diagrammatic systems relied on the use of enclosed curves as the primary device for representation. These tended to be augmented with line segments, points, letter labels, shadings, etc. Some turned out to be merely explorations of the potential of logical diagramming; others turned out to be rich and powerful in terms of representation and deduction. Some turned out to be less so but became popular because of the ease of construction and utility. We also saw that some researcher, such as Leibniz and Lambert, conceived of systems of logic diagrams that relied on the use of line segments as the primary device for representation. However, their efforts did not result in fully worked-out diagrammatic systems and had limited influence on subsequent researches into logic diagrams. Nonetheless, the possibility of constructing a system of diagrams for logic that focuses primarily on the use of line segments rather than closed curves has not been forgotten. Simultaneously with the heterogeneous logic project, and in virtually every way compatible with it, there have been developments in linear diagramming that aim to exploit such systems, to demonstrate their advantages over standard symbolic systems, and to show how such systems compare with other kinds of diagrammatic systems in terms of expressive and deductive power, perspicuity, simplicity, and naturalness. In the next chapter I will introduce some of this work and then offer an extensive presentation of my own work on linear diagrams.

3 Lines of Reason

3.1 New Lines (Smyth and Pagnan)

> [T]he syllogism admits of a graphical representation which is as suggestive as a diagram of geometry.
>
> De Morgan

We saw above that, as Wesoły's research suggests, Aristotle might well have used some sort of diagrams using labelled line segments to illustrate the forms of perfect syllogisms. What is certain is that by the 17th century logicians had begun projects designed to use diagrams not only to reveal the forms of syllogisms but to provide illustrations of syllogistic reductions. It appears that Leibniz (followed then by Lambert) was the first logician to attempt such a system of diagrams for syllogisms based on line segments rather than closed plane figures such as circles. Yet, in the long run, systems of logical diagrams based on the use of closed figures (such as those of Euler, Venn, and Peirce) won out. As we have seen, there was a significant decline in interest in logic diagrams during the first several decades of the 20th century. Nonetheless, we have also seen that there has been something of a "renaissance of diagrammatology" of late. Not surprisingly, attempts to develop systems of logic diagrams based on the use of line segments have been part of that renaissance. One might consider M. B. Smyth's "A Diagrammatic Treatment of Syllogistic" (Smyth 1971) as an early setting of the stage for this part of the renaissance.

Smyth proposed a system of "directed graphs" that could effectively be used to represent standard categorical propositions, with the restriction that they contain no complex, negative, or empty terms (Smyth 1971, 483). The use of such a system allows one to "read off" from a graph of an arbitrary finite set of such propositions all logical consequences of that set. Soundness and completeness proofs for the system are briefly sketched (Smyth 1971, 485). Smyth also explores what he terms "the general structure of valid syllogisms" (Smyth 1971, 485–488). So, what is a directed graph?

A directed graph is a diagram in which each premise of an argument is represented. Each term of a proposition is represented by a point ("vertex") at one end of a line segment that may or may not contain a directional arrow, an interruption, or another term. This is best understood by looking at how the standard categoricals are represented.

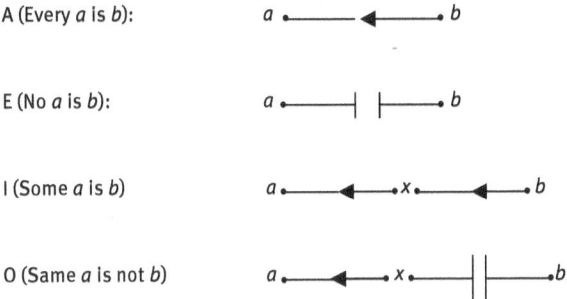

A (Every *a* is *b*):

E (No *a* is *b*):

I (Some *a* is *b*)

O (Same *a* is not *b*)

Figure 3.1: Directed Graphs of A, E, I, and O

It's obvious that Smyth's notational and graphic system is hardly simple. And there is more. In diagramming an argument, Smyth was intent on guaranteeing that the directed graph generally conforms to a shape that has a number of distinct "branches." This is achieved by representing terms that are positively connected to one term (i.e. by line segment containing a directional arrow) but is negatively connected to another term (i.e. by an interrupted line segment) as a line segment that is *bent* from the horizontal at its midpoint. Such terms are labelled on a graph at their midpoints. An example of a directed graph from Smyth (Smyth 1971, 484) illustrates this.

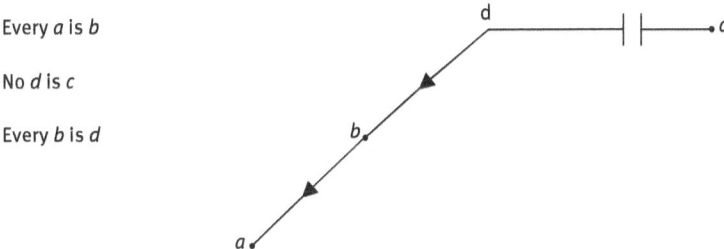

Every *a* is *b*

No *d* is *c*

Every *b* is *d*

Figure 3.2: Directed Graph of an Argument

The conclusions that can logically be drawn here are those propositions that can be read off. How? A vertex is a *descendant* of another vertex if there is a progression of directed line segments leading from the latter to the former. Every vertex is a descendant of itself. Two terms that are on interrupted branches of a graph are mutually excluded.

(i) If *a* is a descendant of *b* in a graph, then A*ab* can be inferred.
(ii) If *a* and *b* are mutually excluded in a graph, then E*ab* can be inferred.
(iii) If *a* and *b* have a common descendant, then I*ab* can be inferred.

(iv) If *a* has a descendant that is excluded from *b* in a graph, then O*ab* can be inferred.

So, the conclusions that can readily be seen to have already been diagramed (thus read off) are those that just happen to be determined by the above rules. In the case of the three premises diagrammed in Figure 3.2, one can initially draw the following conclusions:

A*ad*, by (i)
E*ac*, by (ii)
E*bd*, by (ii)
and, trivially: A*aa*, A*bb*, A*cc*, A*dd*, E*ca*, E*db*.

It will be useful later on in this chapter to make comparisons between Smyth's directed graphs and other similar linear diagram systems. To that end, consider for now directed graphs for the four Aristotelian perfect syllogisms.

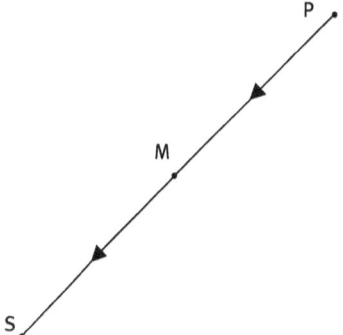

Figure 3.3: Directed Graph of Barbara

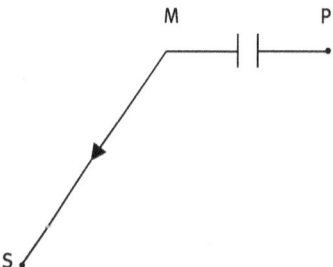

Figure 3.4: Directed Graph of Celarent

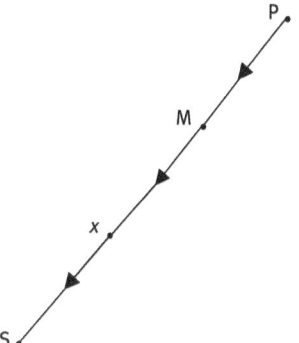

Figure 3.5: Directed Graph of Darii

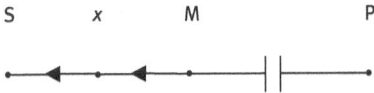

Figure 3.6: Directed Graph of Ferio

The following should be noted. The use of what Smyth calls "arcs" (line segments containing directional arrows) is never clear. Thus, the graphic representation of 'Some a is b' looks as if it could be read as 'Every a is x and every x is b' unless the role of the unknown singular term, x, is more fully specified. Also, in considering directed graphs, the limits imposed on them should be kept in mind: no negative terms, no empty terms, and no complex (thus no relational) terms.

In the period following Leibniz and Lambert, it would seem that Smyth was a pioneer in the area of linear logical diagrams. His work appeared only in a brief essay in 1971. It would be two more decades before another attempt at a system of diagrams for logic based on the use of points and line segments (i.e. linear diagrams). That system was first presented in 1992, and, like Smyth's essay, this one also appeared in the *Notre Dame Journal of Formal Logic*. The next several sections of this chapter will focus on that system. But, first, for its value in its own right and its value for comparisons, we will jump ahead to the 21st century and look at the system of linear diagrams developed by Ruggero Pagnan.

In recent years, Pagnan has been engaged in a program that is rich in promise (Pagnan 2010, 2012, 2013a, 2013b). The heart of this program is a system of linear logical diagrams that is meant to be at once algebraic and graphic. The result is a heterogeneous formal system of logic. We offer below a brief summary of Pagnan's system, highlighting some of its main elements, and provide example diagrams to illustrate it. There is of course more to his program than this summary of his system suggests. For example, he has shown (Pagnan 2013b) how his account of syllogistic can be seen as a part of an intuitionistic version of the sequent calculus of linear logic, and he extends the account, given in earlier papers, of De Morgan's syllogistic with "complemented" (negated) terms. In each of his papers (especially Pagnan 2010), with due credit to Smyth, who had done the same for his own system, Pagnan illustrates how his system can extend to n-term syllogisms. However, it is his core system for diagramming syllogisms (admitting negated terms) that requires our attention now.

Pagnan calls his system "SYLL" a system that is heterogeneous in that it makes use of both graphic and linguistic syntactical objects. Terms are represented by letter labels (called "term-variables") and their relations in a given statement by a series of directed arrows (→,←) and "bullets" (dots: ●). Its diagrammatic representations are linear, making no use of any closed curves. In effect, a well-formed SYLL diagram of a categorical statement is a pair of term letters separated by a finite list of arrows, themselves separated by a bullet or a term letter. The four standard categoricals are represented as follows:

A: S → P,
E: S → ● ← P,
I: S ← ● → P.
O: S ← ● → ● P.

According to De Morgan, "In the form of the proposition, the copula is made as abstract as the terms: or is considered as obeying only those conditions which are necessary to inference" (De Morgan 1966, ix). As Pagnan notes, the series of arrows and bullets between such pairs of terms amounts to an *abstract logical copula*, (Pagnan 2012, 35). One could read the categorical diagrams as, correspondingly, 'P belongs to every S', 'not-P belongs to every S', 'P belongs to some thing that S belongs to', and 'not-P belongs to some thing that S belongs to'. Pagnan also permits statements to be equivalently represented by the "reversals" (mirror images) of their usual diagrams, making use of either kind of representation throughout Pagnan 2012 and 2013b. Thus an A categorical could be represented as: P ← S, so that the arrow here could be read as 'belongs to every'. Note that, reversed or not, the bullets – and their positions relative to the directions of the adjacent arrows – carry substantial logical weight. For example, → ● X (or its reversal) can be taken (though Pagan doesn't say so) as diagramming 'not-X'; a bullet between the tails of two arrows could be taken as the representation of a particular (existential) quantifier; and X → ... (or its reversal) can be taken as diagramming '... belongs to every'.

In SYLL, syllogistic inference amounts to concatenation and reduction. "Two or more syllogistic diagrams can be concatenated and reduced, if possible, by formally composing two or more consecutive and accordingly oriented arrow symbols separated by a single term-variable, thus deleting it" (Pagnan 2013a, 36). If such a concatenation and reduction is possible, then the syllogism is valid; otherwise it is invalid. *Concatenation* and *reduction* works, essentially as follows. The two premises are diagrammed and placed adjacent to each other in such a way that the term common to each (viz., the middle term) is the right-most term of the premise on the left and the left-most term of the premise on the right. Then the two tokens of the middle term are amalgamated into a single token, resulting in a new, single diagram. Finally, the middle term is deleted representing the conclusion and pairs of commonly directed arrows are amalgamated. Often, in practice, the first step is avoided by simply diagramming the premises together while amalgamating the two middle term tokens. For illustration, here are the syllogistic diagrams of the four perfect syllogisms.

Barbara	Celarent	Darii	Ferio
S → M → P	S → M → ● ← P	S ← ● → M → P	S ← ● → M → ● ← P
———	———	———	———
S → P	S → ● ← P	S ← ● → P	S ← ● → ● P

Figure 3.7: SYLL Diagrams of the Four Perfect Syllogisms

Imperfect syllogisms can be given similar graphic treatment, reducing each of them to a perfect syllogism by applying rules of conversion, premise re-ordering, and term substitution. In each case, a necessary (but not sufficient) condition for the validity of any diagrammed syllogism is that the diagram of the conclusion contains as many bullets as the conclusion (Pagnan 2013a, 36).

So-called strengthened (sometimes weakened) syllogisms can be treated in SYLL as well. According to Pagnan, existential import is not implicit in any categorical other than one of the form 'Some X is X' ('Something is X', 'There is an X', 'Some thing is X', 'There exists an X', etc.). This is always the form of a contingent statement, and the diagram for which is X ← ● → X. Its contradictory, 'No X is X must be contingent as well. Moreover, since 'Some X is not X' is a contradiction, 'Every X is X' must be tautologous. Pagnan calls X → X and X ← ● → X the "laws of identity" (Pagnan 2013a, 40). Strengthened syllogisms require an implicit premise of this form. Consider the fourth figure syllogism Baramtip.

The two universal premises must be supplemented by an implicit premise of the form 'Some S is S'. Diagrammed in SYLL, this yields:

S ← ● → S → M → P

_____

S ← ● → P

Figure 3.8: SYLL Diagram of Baramtip

Note the new rule applied here. Two tokens of a given term letter flanking ← ● → can be amalgamated. Thus ... X ← ● → X.. simply reduces to ... X ... ..

Indirect reductions, as usual, assume the contradictory of the conclusion and then derive an explicit contradiction. Such a contradiction will have the form X ← ● → ● ← X. For example, here is a diagrammatic proof that an E statement and the corresponding I statement are contradictory by deriving a contradiction from the pair of them:

S ← ● → P   P → ● ← S

_____

S ← ● → ● ← S

Figure 3.9: SYLL Proof that E and I are Contradictory

Note that in this derivation that 1) the E premise is used in its equivalent revered form and that 2) the conclusion has the form of a contradiction, viz. X ← ● → ● ← X.

One of the ways Pagnan wants to extend traditional syllogistic is by incorporating "complemented" (negated) terms (Pagnan 2013a, 49), which he introduces in the context of his partial adoption of De Morgan's *spicular notation*. De Morgan used the lowercase form to represent a term's negation (thus 'a' is the negation of 'A'). This would allow a slight simplification in Pagnan's formalism. An E categorical could now be represented as S → p and an O categorical could be represented as S ← ● → p. De Morgan's spicular notation uses left and right parentheses to symbolize quantity and dots to indicate affirmation or denial. In the spicular system of logical notation, X) and (X are meant to formalize 'every X' while X(and)X formalize 'some X'; an even (or no) string of dots between parentheses (whatever the orientation of the parentheses) indicates affirmation and an odd string of dots indicates denial. For example, an E categorical would be symbolized as S)●(P and an O categorical would become S(●(P. Using De Morgan's notation for term negation, the four standard categoricals can be more simply represented as follows;

A: S))P
E: S))p
I: S( )P
O: S( )p

Pagnan claims that his diagrammatic system supports De Morgan's formal system:

> The possibility of making a distinction between a term being universally or particularly quantified, as well as between affirmative and negative modes of predication is supported by the diagrammatic formalism [SYLL] ... together with the possibility of handling complements of terms. Indeed, we can look at the symbols ●, → and ← as to fundamental ones. In a diagram built as a combination of such fundamental symbols, a term-variable X is universally quantified if it enters in it as X → or ← X, whereas it is particularly quantified if it enters in it as X ← or → X. The complement of X is represented as X → ● or ● ← X, both of which may be abbreviated as x. (Pagnan 2013a, 50)

For the most part, Pagnan is right here. We have already noted how the right combination of arrows and bullets can represent term negation. But Pagnan (and sometimes De Morgan) is wrong in thinking that a combination of an arrow adjacent to a term always indicates quantity, for this would commit him to the quantification of predicate-terms. For example, the universal affirmation, X → Y, would have to be interpreted as 'Every X is some Y'. What Pagnan's arrows (and De Morgan's parentheses) actually indicate is the *distribution value* (distributed or undistributed) of the term to which they are attached. Put simply, terms at the tails of arrows are distributed; terms at the heads of arrows are undistributed. In order to see this more fully one must get clear about both distribution and the notion of a logical copula in a logic of terms (such as syllogistic). These things will come under much closer scrutiny in subsequent sections of this chapter. But, before going on to those sections, a few general comments and observations about SYLL are in order now.

The SYLL system guarantees that all diagrams are unambiguous. It can accommodate n-term syllogisms. It represents the symmetry of E and I form statements in an explicit and obvious way. SYLL accurately takes a strengthened syllogism to be an enthymeme whose tacit premise has the form 'Some X is X',

'There is an X', 'An X exists', etc. (X ← ● → X). SYLL dispenses with the traditional subject/predicate distinction. As well, SYLL is able to represent term complementation (term negation). Finally, SYLL implicitly (but unadmitted by Pagnan) is able to represent the distinction between distributed and undistributed terms. These are all important characteristics that are wanted in any viable formal account of syllogistic logic. Nonetheless, SYLL does have some unwelcome characteristics. SYLL provides no means of symbolically or graphically representing the contradictoriness of a pair of contradictory statements – no way to represent sentential negation. It provides no way to represent relational terms. Furthermore, it makes no provisions for the representation of singular terms (e.g., proper names, definite descriptions, anaphoric pronouns, etc.). Finally, there is a feature that is most disturbing about SYLL. As we have seen, the difference between linguistic expression and graphic representation is a matter of degree rather than kind. Pagnan's diagrams seem to fall on the scale much closer to the linguistic end than to the graphic end. They look more like logical formulae than logical diagrams. S ← ● → P looks closer to $\exists x(Sx \& Px)$ than to Euler's two overlapping circles. Indeed, the string of symbols ← ● → looks to be simply a symbolic rendering of Aristotle's 'belongs to some'. Pagnan's diagrams *are linear*. But are they *diagrams*? SYLL "diagrams" certainly don't appear to be graphic in any significant sense. SYLL proofs don't seem to rely on visual inference. SYLL might easily be construed as nothing more than an alternative symbolic system, on a par with the standard symbolic system of first-order monadic predicate logic.

So what *is* wanted in a linear (and two-dimensional) diagrammatic system for logic? In the remainder of this chapter we will present such a system of logical diagrams, one that will go well beyond previous systems in its graphic capacities. It is the system referred to earlier, the one first presented in 1992a. It is the system called *Englebretsen Diagrams* (Rauf 1996, 397–408).

3.2 In Logical Terms: Term Functor Logic

> All our logics are now but a shadow of what I should wish and what I see from afar.
> Leibniz

When Venn developed his system of logical diagrams, he did so with the intention of providing a graphic analogue to Boole's algebra of logic. The system of Englebretsen Diagrams (from now on (ED) (Englebretsen 1992a, 1996) was likewise meant to provide a graphic analogue to Fred Sommers' term logic (Sommers 1967, 1969. 1970, 1973, 1975. 1976a, 1976b, 1976c, 1976d, 1981, 1982, 1983a, 1990,

1993, 2000, 2005a, 2005b, and Sommers and Englebretsen 2000). Sommers' version of term logic, which he came to refer to as Term Functor Logic (TFL), was explored, defended, emended, and exploited in many places by Englebretsen (for example, Englebretsen 1981, 1996.2002, 2005, 2015, and especially 2016b; also Englebretsen and Sayward 2011). The key to understanding any term logic such as Aristotle's syllogistic logic or Sommers' TFL is its account of logical syntax.

The grammatical form of any natural language statement is determined by the grammatical conventions of the pertinent linguistic community. Such conventions (grammar rules) exhibit a variety of features that are meant to help encode information. Different languages make do with featuring different kinds of information. Thus some languages exhibit a variety of tenses, others systematically encode gender, and so forth. As well, natural languages often exhibit redundancy of information (for example, English marks number (singular/plural) on both subject expressions and main verbs). Much of this sort of information is immaterial to the interests of logicians. The logician seeks those features of a statement that are involved in determining the statement's role in logical inference, etc. In short, the logician looks beneath the "surface" grammar of natural language statements in hopes of finding their *logical grammar*, their logical forms. When Aristotle sought logical form, he initially followed his teacher Plato by taking his clues about what the logical form of a statement might be from select features of his native Greek. Plato and the early Aristotle took the basic logical form of any statement to be a noun connected to a verb (e. g., 'Theatetus walks'). A logically formed statement cannot be constructed from a pair of nouns, nor can it be formed from a pair of verbs. One noun and one verb are required. The logical form is the grammatical form minus any accidental features that are of no logical concern. How does a noun connect with a verb to form a statement rather than just a pair of words? How is unity achieved? According to Plato, the noun and the verb "mix" in the sense that they are simply *fit* for one another. Think of the verb as a board with a round hole and the noun as a round peg fitting that hole. As it happens, two-and-a-half millennia later, that turned out to be just the idea Frege had when he came to the problem of determining logical form and the so-called "problem of propositional unity." Of course Frege took his clues about logical form not from Greek grammar (or even the grammar of his German). His clues came from the mathematical notion of functions. A function applies to one or more "arguments" to yield a new expression that has a value. For example, the square root function applied to the argument 4, yielding an expression ($\sqrt{4}$), has the value ± 2. For Frege, the logical form of a simple statement was a function (essentially a predicate), which is "unsaturated" or incomplete (having holes) along with an appropriate number of arguments, "names" (essentially singular denoting expressions) that are "saturated" or complete and fitting into those holes. The result is a statement whose logical unity is guaranteed by the pegs filling the holes so that the statement is complete. Thus a statement is, in turn, itself a peg, an expression fit for filling the holes in "higher" predicates (e..g., sentential "truth functions" such as 'if...then' and 'or'). As well, a statement, being the result of a function applied to an appropriate number of arguments, has a value. For Frege the value of a statement is its truth-value (what Frege called "the True" and "the False"). Theories of logical syntax such as these require that statements be construed as unified strings of terms ... but there must be two fundamentally distinct kinds of terms: nouns and verbs (or names and predicates).The logician was thereby committed to a heterogeneous logical vocabulary, lexicon. But Aristotle eventually abandoned any such a theory.

By the time he composed *Prior Analytics*, Aristotle realized that mediate inference must involve at least two premises that share a term in common. Consequently, at least three terms are involved in such inferences. Moreover, and most importantly, Aristotle saw that at least one of those terms had to play a different logical role in at least two different statements. Since a noun can never play the role of a verb, nor can a verb ever play the role of a noun (from a logical point of view, they are fundamentally distinct kinds of terms), logical form cannot rest on such a distinction (see Englebretsen 1982b, 1986b). Aristotle's solution was to adopt a logical lexicon that is homogenous; it simply consists of *terms*. Those terms certainly have grammatical features and grammatical roles to play, but these are beside the point of logic. Any term can play any logical role that any other term can play. So, how is a statement more than just a sting of terms? How is it a unit – logically? Here is an example of Aristotle's genius. Pairs of terms are bound together to form a statement, a unit fit for the role of premise or conclusion, by a *logical copula*. A logical copula binds pairs of terms together. It literally facilitates the copulation of pairs of terms. A logical copula is a special-duty expression (a *formative*, a *syncategorematic* expression, a *logical constant*) whose job in a statement is manifold: it unites the terms, it determines the *quality* of the statement, and it determines the *quantity* of the statement. And it does this all at once. This is Aristotle's theory of logical syntax. It is what makes syllogistic logic possible. On this theory of logical syntax, a statement is not the result of nouns/names fitting verbs/predicates; it's a product of pairs of expressions (terms) being bound together: not pegs fitting holes but blocks glued together.

For Aristotle there were four kinds of glue, English versions of which are: 'belongs to some', 'belongs to every', 'belongs to no', and 'does not belong to some'. Statements made by flanking any of these with a pair of terms are *cate-

gorical. Thus, for example: 'Humour belongs to some speakers', 'Reason belongs to every man', 'Reason belongs to no fish', 'Humour does not belong to some speakers'. Syllogistic logic is the logic of categoricals. Notice that these English paraphrases are understandable but awkward (that was also true for Aristotle's Greek paraphrases). Aristotle could have eased this awkwardness by not merely paraphrasing but by fully *symbolizing*, rendering categoricals in an artificial language meant to display the logical form free of any natural language expressions. Aristotle did not go that far. However, he did take an important step in that direction. He allowed letters to stand in the place of natural language terms. In other words, he introduced symbolic *variables*, such as A, B, Γ, etc. to stand for different terms in different logical contexts. To this degree, Aristotle's syllogistic logic was a *symbolic logic*. Term variables certainly relieve some of the awkwardness of the categorical paraphrases ('H belongs to some S' is marginally better than 'Humour belongs to some speakers'), but not much. Late medieval logicians helped here. They *split* the copula: one fragment went with one term and another fragment went with the second term; Next, they rearranged these terms and fragments so that they look more like natural language statements (a kind of logicized Latin (for a brilliant account of such a language, which he calls "Linguish," see Parsons 2014). Thus, for example, 'H belongs to some S' became 'Some S is H'. This splitting of the copula reveals two of its functions in the statement by assigning each to a different fragment. Here, 'some' indicates quantity and 'is' indicates quality (but see Englebretsen 1990b and 1997). However, it is important to remember that quantity and quality do not characterize the terms to which the *quantifier* and *qualifier* are assigned. Quantity and quality are features of the categorical statement as a whole. The copula is split into two fragments – but they are just that, fragments of a single quantifier. Traditional logicians countenanced both split and unsplit versions of the copulae. Consequently, categoricals with their copulae unsplit could be rendered fully symbolic by allowing the copula to be symbolized using lowercase letters corresponding to the A, E, I, O that labeled categorical forms in general. Thus: SaP, SeP, SiP, SoP.

Term Functor Logic, TFL, rests on a theory of logical syntax that amounts to the traditional theory ... up to a point. Traditional syllogistic logic could not easily accommodate three kinds of expressions: singular terms, complex (including relational) terms, and compound (truth functional) statements. Aristotle was well aware of all of this. He may have had reasons to think that singular statements are rarely encountered in science, but in both *Analytics* one can find many examples of them (*Prior Analytics* 43a34–35, 47b24–25, 47b32–33, 67a33–37, 68b41ff, 69a2, 69b12–15, and *Posterior Analytics* 78b4–10, 90a5–25, 93a30–33, 93a36–67) (see Englebretsen 1980b). Moreover, there is good reason to believe that Aristotle understood that just as singular terms can be predicated in natural language, the same holds for the logical forms of such sentences (see *Topics* 15234 ff). One consequence of this is that syllogistic requires no special account of "identity statements" (Sommers 1967, 1969, 1976c, 1976d, 1982, 1990, 2000b, Englebetsen 1982a, 1985f, 1996, 2015). Post-Fregean logic takes great pride in the notion that it can easily account for the logic of statements involving relational terms, marking its primary advantage over traditional logic. Yet Aristotle, though unable to fully incorporate the logic of relationals into his syllogistic, was at least aware that a way to do so was required. He made a start by providing examples of inferences involving relationals and attempted to formulate rules governing them (*Topics* 114a18–19, 114b40–115a1–2, 119b3–4, and *Prior Analytics* 48b11–24; see also Bocheński 1968, 68–69). Yet, he was never able to see how relational statements contain more than one referential term (viz. subject- and object-terms) (see Thom 1977 and Englebretsen 1982d). Finally, one should not be surprised to learn that Aristotle was confident that he could account for the logic of so-called truth-functional statements. He promised to carry out the task but apparently never did. Aristotle gave examples of inferences involving such unanalyzed (into terms) statements (*Prior Analytics* 53b12–24, *Posterior Analytics* 57a36–37, 75a2–4; see also Bocheński 1968, 70–71). What was missing was the idea that entire sentences can be treated as (complex) terms, which allows the logic of statements to be treated as merely a special branch of a logic of terms (see Sommers 1993 and Englebretsen 1980c).

Medieval logicians made some progress with all three of these challenges (Parsons 2015). But it wasn't until Leibniz's efforts along these lines that real progress was made. Medieval logicians tended to construe singular statements as implicitly universal in quantity, thus fit for roles as premises or conclusions of syllogisms. However, Leibniz realized that the quantity of such statements could be either universal or particular, depending on the logical environment in which they were being used (Leibniz 1966, 115). As we will see, this was far from a minor adjustment that strengthened syllogistic logic. Leibniz also saw that a relational expression, a relational term along with its object term (e.g., 'loves a philosopher') can be taken logically as a single complex term. This allowed him to analyze inferences involving such relationals as straightforward syllogisms. He actually provided a proof of the inference 'Painting is an art; therefore, he who learns painting learns an art' (Leibniz 1966, 88–89). Three centuries later, De Morgan dealt with a now more famous version of an inference of the same logical form. Finally, and most importantly, Leibniz offered valuable insights into how the logic of compound statements (e.g., conjunctions, disjunctions, conditionals), the so-called "hypothetical" statements, could be incorporated into the logic of terms (see especially Castañeda 1976). He saw that entire *statements* can be construed as *terms*. "[T]he categorical proposition is the basis of the rest, and modal, hypothetical, disjunctive and all other propositions presuppose it" (Leibniz 1966, 16). "[A]bsolute and hypothetical truths have one and the same laws and are contained in the same general theorems, so all syllogisms become categorical" (Leibniz 1966, 78).

> If, as I hope, I can conceive all propositions as terms, and hypotheticals as categorical, and if I can treat all propositions universally, this promises a wonderful ease in my symbolism and analysis of concepts, and will be a discovery of the greatest importance. (Leibniz 1966, 66)

Unfortunately (as with his systems of logical diagrams), all these Leibnizian insights were generally unknown for a very long time. Some of them were independently acquired by others in the 19th and 20th centuries (e.g., the treatment of relationals by De Morgan and Peirce). Eventually all of them were more clearly articulated and strengthened in Sommers' development of TFL (Sommers 1976a, 1976b, 1982, 1993, 2000, Englebretsen 1981, 1982c, 1985a, 1987a, 1988, 1996, 2015). Before continuing, it should be noted that Sommers' TFL was not the only system of term logic developed in the 20th century. For it turns out, perhaps ironically for some, that one of the most prominent champions of MPL, Quine, also formulated a version of term logic, Predicate Functor Algebra (PFA) (see Quine 1936a, 1936b, 1937,1959, 1960a, 1971, 1976a, 1976b, 1981a, 1981b and Noah 1980, 1982, 1987, 1993, 2005; see also Bacon 1985). Of course, he did so only to highlight the crucial role of individual variables and the quantifiers that bind them in MPL by eliminating them. Once eliminated, their roles had to be assigned to new formal elements, "predicate functors." which recursively apply to predicates to form new predicates. The result was a logical syntax admitting only predicates (which, since they are now the only kind of non-formative expression, might just as well be called *terms*) and functions on them. Once rules for the logical manipulation of formulas in the new language are provided, the result is PFA.

Quine built his version of term logic in order to highlight certain fundamental features of standard predicate logic. Sommers, by contrast, built *his* version of term logic (TFL) in order to reveal the logic of natural language. Consequently, TFL's system of symbolization is simple, natural, and perspicuous. It preserves a number of old (but now often forgotten) logical insights. The system of symbolization for TFL rests on Aristotle's notion of logical syntax in terms of copulated term pairs; it admits the medieval practice of splitting copulae; it borrows and expands on Leibniz's insight that singular terms, when in subject or object roles, have arbitrarily universal or particular quantity ("wild" quantity in Sommers 1969 and Englebretsen 1986a, 1988); it refines De Morgan's idea that the logical copula should be as abstract as possible (De Morgan 1850 and 1966, ix, 51).

TFL also incorporates the Leibnizian view, noted above, that the logic of compound statements (truth-functional logic) can be construed as a part of the logic of terms (Leibniz 1966, 16, 66, 78). Finally, and crucially, TFL entrenches the twin theses that terms, logically, come in oppositely charged pairs and that logical formatives are signs of opposition. Thus:

> To reason therefore is the same as *to add* or *to subtract*, ... Therefore, all reasoning reduces to these two questions of the mind, *addition* and *subtraction*.(Hobbes 1981, 177)
>
> So just as there are two primary signs of algebra and analytics, + and –, in the same way there are as it were two copula, 'is' and 'is not'... (Leibniz 1966, 3)
>
> I think it reasonably probable that the advance of symbolic logic will lead to a calculus of opposite relations, for mere inference, as general as that of + and – in algebra. (De Morgan 1966, 26)
>
> We shall take this suggestion of Leibniz quite seriously, and see where it leads. (Sommers 1976a, 20)
>
> Both terms and propositions come in opposed pairs. Opposed terms are called logical contraries. ... Opposed propositions are called contradictories. (Sommers 1982, 169)
>
> It is a surprising but little known fact that familiar logical words we use in quotidian deductive reasoning behave in a natural language like English in just the way that '+' and '–' signs behave in algebra and arithmetic. (Sommers 2005a, 59)

The formal symbolic language of TFL consists of countably many uppercase letters that are variables for natural language terms, the two functors (syncategoremata) + and –, and parentheses pairs as needed for grouping. Terms always come in oppositional pairs (e.g., 'happy'/'unhappy', 'massive'/'massless', 'colored'/'colorless', 'in the car'/'not in the car'). Term letters, then, are always logically charged positively or negatively (e.g., $+H/-H$, $+M/-M$). In such cases, the signs of term charge are unary, applying to one term. But the same plus and minus signs can be binary (applying to a pair of terms) as well. Such binary signs are either unsplit logical copulae or the fragments generated by a split copula. For example, Aristotle's standard I and O categoricals could be paraphrased as 'P belongs to some S', and 'nonP belongs to some S', and symbolized as 'P+S' and '–P+S'. Note that in this symbolization for O the '–' is unary and the '+' in each is binary (an unsplit copula). Since A and E are the contradictories of O and I, these latter can be negated to yield '–(–P+S)' and '–(P+S)'. Note that the result of copulating a pair of terms is a new (more complex) term, a term which itself can be negated (as when the particular forms are negated to yield the new universal forms. Also, notice that, just as in algebra and arithmetic (not to mention natural language) unary pluses are normally left tacit unless explicit use is needed. Moreover, again as in algebra and arithmetic, the plus and minus signs are systematically ambiguous between their unary and binary uses (compare positive and negative numerical expressions, on the one hand, with addition and subtraction, on the other).

Our versions of the A and E categorical forms look unfamiliar. We could rectify that by algebraically distributing the outside minuses into the complex expression inside the parentheses. The results would be: A: 'P−S' and E: '−P −S'. In these formulae, the first minus in E is unary and the other minuses are binary; they could be read as 'belongs to every'. We could make the universal forms even more natural looking by splitting those negative copula, in effect formulating 'belongs to every' as 'every ... is ...'. Here the 'every' is a (universal) quantifier and the 'is' is a positive qualifier. So, now, we can apply the splitting procedure to A above to yield '−S+P' and to E above to yield '−S+ −P' ('Every S is nonP'), which, in turn can be taken as '−S−P' ('No S is P'). In statements with split copulae, the *subject* consists of a quantifier and a term (the "subject term") and the *predicate* consists of the qualifier and a term (the "predicate term"). No term, by itself is either a subject or a predicate. Singular terms are terms understood as denoting just one individual (e.g., proper names, definite descriptions, anaphoric singular pronouns). When a singular term is a subject term it arbitrarily admits either a universal or a particular quantity. If, in such a case, the appropriate quantity is either undetermined or logically unimportant, the quantity is symbolized by '*'. Thus 'Socrates is wise' would be symbolized as '*S+W'.

Any pair of terms can be conjoined to form a new complex term. Such conjunction is affected by either the split or the unsplit positive copula. For example, 'rich and famous' is symbolized with the unsplit + as 'R+F' and its equivalent 'both rich and famous' is symbolized with the split copula as '+R+F'. Since *any* term can be conjoined with any other term to form a complex term, singular terms and compound terms can be conjoined with any other term to form a complex term. Examples are 'Tom and Jerry', 'rich and famous but unhappy'.

Relational terms are terms; in fact, they are compound terms with unsplit copulae. Consider 'Plato teaches some mathematicians'. It has three terms, but every statement is, from our logical point of view, a pair of copulated terms (i.e., a complex term). Here, 'Plato' is the subject term. So we can make a start at symbolization with '*P teaches some mathematicians'. Now 'teaches some mathematicians' is a relational expression, consisting of a *relative term*, 'teaches' and 'some mathematicians'. The entire relational expression is a complex term consisting of a pair of terms joined by an unsplit copula, which happens to be, in this case, 'some' (as in 'belongs to some'). The full symbolization would be '*P+(T+M)'. When a relational has another relational as one of its terms, as in 'Every candidate made some promises to every voter', it can just as easily be symbolized. Thus, in this case: '−C+((M+P)−V), in which the relational term is itself a relational term.

Entire propositions are complex terms. They are the results of copulating pairs of terms (which themselves might be general terms, singular terms, conjoined terms, relational terms, or even propositions). Consider 'Some philosophers are logicians and every logician is rational'. Its two conjuncts can easily be symbolized as '+P+L' and '−L+R'. Moreover, we know we can take the 'and' here as an unsplit copula to form the conjunction: '(+P+L)+(−L+R)'. Suppose our logical context does not require that propositions be analyzed into their sub-sentential terms. Let us use lowercase letters (e.g., p, q, r, ...) as letter variables for them. Thus 'p+q' for the above example. Since 'p' and 'q' stand for unanalyzed propositional terms (compound terms) they have a charge. In this case the charge for each is positive (so tacit). Such terms could have negative charge. Thus '−p' would be the negation (viz., contradictory) of 'p'. Since we can reformulate any complex of propositions (conditionals, disjunctions, etc.) entirely in terms of negation and conjunction, we can symbolize all such truth-functionals using only propositional variables along with appropriate complements of pluses and minuses. Thus, since 'p+ −q' is the contradictory negation of '−(p+ −q)', which is the form of 'Not both p and not-q', which in turn is equivalent to 'If p then q', we might well symbolize the latter as '−(p+ −q)'. Better still would be to drive in the external minus to give us '−p+q', taking this to be the most appropriate symbolization for 'If p then q'. We can deal with disjunctions (e.g., 'p or q') in a similar manner to yield '−(−p + −q)' or '− −p − −q'.

Notice that the symbolization of 'If p then q' (−p+q) shares the same form as the universal affirmation 'Every S is P' (−S+P). Indeed, the same holds for all of the four standard categorical forms. 'Some S is P' (+S+P) and 'Both p and q' (+p +q); 'No S is P' (−S−P) and 'Neither p nor q' (−p−q); 'Some S are not P' (+S−P) and 'p but not q' (+p−q). In fact, as Leibniz, Kant, Boole, Peirce, and others have noticed, the logic of terms and the logic of unanalyzed propositions are either isomorphic, or identical, or the latter is a "special branch" of the former (Sommers 1993). It is only the purely formal features (reflexivity, symmetry, transitivity) of logical copulae (whether applied to pairs of terms or pairs of propositions) that are of logical import. It seems De Morgan saw this when he wrote that the copula should be construed as abstractly as possible.

One might object to the symbolic system of TFL by pointing to the ambiguity of the plus (+) and minus (−) signs. They are allowed to play two different roles in the syntax: they are marks of term charge (e.g., +A/−A) and they are fragments of split copulae (in effect, marks of quantity and of quality). Now, while ambiguity is often an obstacle to clear and efficient communication in natural language, ambiguity can be a source of expressive power. This is clearly seen in the use of formal languages such as in mathematics. Consider Arabic numeration (compared with Roman numeration). The expression '222' makes use of three instances of the numeral '2'. But in each instance the position of the numeral signals that it is to be understood differently from the other two tokens of the numeral. The first is understood as '2×10^2, the second as 2×10^1, and the third as 2×10^0, i.e., 200 + 20 + 2. And there is more. The simple plus and minus signs are systematically ambiguous. They are allowed to play two different roles: they are marks indicating whether a numerical expression is positive or negative (e.g., +42/−42) and they indicate the operations of addition and subtraction. In other words, these pluses and minuses are sometimes unary operators (applied to single numerical expressions) and sometimes binary operators (applied to pairs of such expressions). This kind of ambiguity is a very good thing in mathematics. It is also a very good thing in the symbolic system of TFL.

Needless to say, it's one thing to build a symbolic logic and quite another to build a diagrammatic logic. A symbolic system adopts a number of conventions for interpreting its expressions. Natural languages are symbolic systems. Artificial systems of fully symbolic formal logic are meant to facilitate the translation (via appropriate intermediate paraphrases when required) of natural language statements into well-formed (grammatically correct) formulas of the system's artificial language. Such symbolization aims to reveal logical form, the "canonical" form according to logicians like Quine. The advantage to be gained by symbolization is the increased transparency of form and ease in the process of carrying out various logical tasks. "If we were to devise a logic of ordinary language for direct use on sentences as they come, we would have to complicate our rules of inference in sundry unilluminating ways" (Quine 1960b, 158). A sufficiently well-constructed *diagrammatic* system (in particular, one that is heterogeneous) enjoys these same advantages. But, it also enjoys the added advantage of graphically *showing* logical forms. Thus, it allows for *visual inference*, the drawing of a conclusion simply by looking at a diagram without further manipulation.

3.3 Term Lines

> Some schemata are visibly verifiable ...
>
> Quine

> As a practical method of appraising syllogism, rules are less convenient than the method of diagrams ... The diagram test is equally available for many arguments which do not fit any of the arbitrarily delimited set of forms know as syllogisms.
>
> Quine

Since Descartes' development of analytic geometry, mathematicians have learned that the system of real numbers is isomorphic with the geometrical line. In 1995, Hillary Putnam (Putnam 1995) showed how Peirce had eschewed this lesson by holding a conception of *line* (and *point*) that shared much in common with that of ancient geometers as well as Aristotle (cf. Roeper 2006; Shapiro and Hellman 2015; Linnebo, Shapiro and Hellman 2016). Here is how Putnam summarizes Aristotle's conceptions of lines and points:

> [T]he Aristotelian view that points are simply conceptual divisions of the line ... the line is an irreducible geometrical object, not a collection of more elementary objects. ... For [Aristotle], points do not *belong* to lines, although they *lie* on them; that is, they are divisions of them (and also terminations of them, in the case of line segments and curves with endpoints). (Putnam 1995, 4–5)

He goes on to say that according to the ancient view, say Aristotle's or Euclid's (as opposed to the modern view),

> the endpoints [of a line segment] are to be regarded, not as *members* of the line segment ... but as loci distinguished by the fact that an object we have constructed or considered *ends there*. ... The endpoints ... are abstract properties of the line segment itself. (Putnam 1995, 5)

Accordingly, one can think of geometrical points as demarcated, *delineated*, literally, de-lined, conceptually abstracted, from line segments. To use Putnam's word, they are "distinguished" by us according to what "we have constructed or considered." Such conceptual abstraction simply sets them into relief for our attention. Thus, geometrical objects are not constructed from more elementary geometrical objects (ultimately, points). Rather, such more elementary geometrical objects are delineated from less elementary geometric objects. Such delineation does not create, bring into being, geometrical objects; it is merely the result of our *focus*, our conceptual attention to what was previously undelineated. Any point to the left of the right terminal point of a term line is *delineatable* (though very few are actually delineat*ed*).

While geometrical objects are abstract, they can be represented graphically. Such representations are just that – representations. They are physical, perceptible objects. Nonetheless, as representations, they *show* various properties of abstract geometrical objects as well as certain relations that hold among those geometrical objects. It is this feature of such representations, diagrams, that makes them so instructive and useful in geometrical reasoning. Indeed, the same can be said for diagrams used in logic. Russell wrote that a good notation is like a live teacher. One could add that a good system of diagrams is just as much like a live teacher.

The ED system makes use of the following *graphic* elements: straight line segments, points, vectors, a rectangular border. Straight line segments represent individuals denoted by a given term. Uppercase term letters are used as straight line segment labels, the charges of which are indicated by + or −. Positive charge is often suppressed. Points represent specified individuals. Lowercase term letters are used as labels for such points and may be labelled as wild in quantity (*). Vectors (directed arrows) represent relations between or among individuals or sets of individuals; they are labelled just like non-relational terms. Rectangles demarcate pertinent domains of discourse (sometimes "worlds"). The following *conventions* apply: the right-most endpoint of a straight line segment is labelled and that label applies to every point on that line segment to its left; any proper part of a straight line segment is a either a straight line segment or a point; any point on a straight line segment can be indicated to represent a specified individual; vectors are labelled by adjacent term letters; vectors share points with at least two line segments. 'Straight line segment' will be abbreviated as 'term line or 'line'. Vectors need not be straight. Demarcation of domains of discourse by use of rectangles will often be suppressed and left as understood unless context requires otherwise. Nonetheless, it must always be understood that any diagrammatic representation is relative to a specifiable domain. This is because the statements we make are themselves always made relative to some specifiable domain. Normally, such domains of discourse consist of things in the actual world or some salient spatially or temporally understood part of that world. But a domain can be a merely possible world or a fictitious world. This notion of domains will be explored more closely when it comes to diagramming unanalyzed statements (propositional logic). Finally, it should always be remembered that only in the context of a diagram do lines, points, and vectors have meaning (i.e., represent anything).

Systems of logical diagrams exploit the fact that the relations that can hold between pairs of sets (inclusion, intersection (overlap), and exclusion) are easily represented by pairs of geometric plane figures that can themselves be arranged into these same relations. Systems such as those devised by Euler and Venn, for example, use closed figures such as circles, for this purpose. Leibniz and Lambert made a start at using open figures – straight line segments. The ED system exploits the use of line segments as well. Any pair of straight line segments can be arranged into the relations of inclusion, overlap, and exclusion. Moreover, given such an arrangement, the relations between what the lines represent can readily be seen. What is true of any viable system of logical diagrams, whether using closed or open plain figures, is that it rests on the analogy between set relations and geometric relations. Peirce expressed this view in many places (e.g., Peirce 1931–1958, 2.277) and perhaps more clearly than anyone when he wrote:

> [A]ll of deductive reasoning ... involves an element of observation; namely, deduction consists in constructing an icon or diagram the relation of whose parts shall present a complete analogy with those parts of the objects of reasoning, of experimenting upon this image in the imagination, and of observing the results so as to discover unnoticed and hidden relations among the parts. (Peirce 1931–1958, 3.363)

Of course, before "observing the results" we need a diagram with "parts" that can stand in relations.

Let T be a term variable. The number of individuals denoted by a given term in any particular context, universe of discourse, is often undetermined. The individuals that constitute the denotation of T are represented by a term line, which, correspondingly, is often undetermined. The term line representing T (viz., T's denotation) is labelled by T placed near its right terminus point:

Figure 3.10: A Term Line

If T is singular, denoting just one individual, it is diagrammed as a single point:

●T

Figure 3.11: A Singular Term Point

A term having no denotation, an empty term, can be diagrammed in two alternative ways: it can be inscribed in the diagram attached to nothing else, or it can simply not appear in the diagram.

Note that the tokens of T above are not accompanied by a sign of positive or negative term charge. But we know that every term does have, from a logical point of view, such a charge. We will assume that any term or term variable not accompanied by such a sign will be understood as implicitly positive (thus T is understood as +T). Needless to say, a negative term, nonT (−T), will be graphically expressed by an appropriately labelled term line.

Figure 3.12: A Negative Term Line

Note as well that a term line represents the entire denotation of the corresponding term. Consequently, one could read a term line such as that in Figure 3.10 as a graphic representation of 'every T'. Likewise, the term line in Figure 3.12 represents 'every nonT' (it does not represent 'no T'.

As noted above, the relations that hold between the lines and points of a diagram (inclusion, intersection, exclusion) are meant to mimic the relations that hold between objects represented by those lines and points. This "relation-based" approach contrasts with the "region-based" approach taken by closed figure diagram system such as those based on Euler or Venn diagrams (for more on this distinction see, for example, Mineshima, Okada, and Takemura 2009, 2010, 2012, and Sato, Mineshima, Takemura 2011). It's time now to show how this is done in ED.

The inclusion of (the extension of) one term in another is simply represented by the line representing the first term being made a (proper) part of the line representing the other. Thus a universal affirmation of the form 'Every S is P' is represented as the following (keeping in mind that the label on a term line is meant to apply to the entirety of the line to its left):

Figure 3.13: Universal Affirmative Line Diagram

Universal affirmations are diagrammed by making one line a proper part of a second. Sometimes, however, two lines will each be (non-proper) parts of one another. This will be so when they are meant to represent a pair of co-extensive terms. For example, to use Quine's well-known example, every creature with a kidney is a creature with a heart and every creature with a heart is a creature with a kidney. This could be diagrammed by giving both terms variable letters that can both label a common line. Let K and H (obviously) be the two labels:

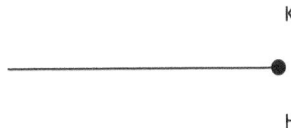

Figure 3.14: Line Diagram for Two Co-extensive Terms

A universal negative ('No S is P') is simply diagrammed with the two labelled lines sharing no point:

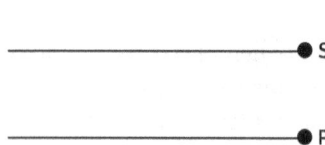

Figure 3.15: Universal Negative Line Diagram

A particular affirmation requires that the two terms of the statement both denote at least one individual in common. The representation of such a statement ('Some S is P') consists of the two lines representing the two terms sharing at least one common point – i.e., intersect:

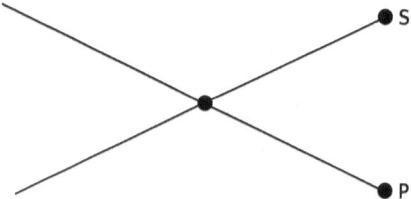

Figure 3.16: Particular Affirmative Line Diagram

As it turns out, there are two alternatives for diagramming particular negative statements ('Some A is not B'). Such a statement may be construed as either (i) claiming that some S is not (= isn't) a P (predicate denial) or (ii) claiming that some S is a nonP (predicate term negation). The first is entailed by the second and is treated so in TFL, (which simply amounts to accepting the traditional rule of *obversion*). The two versions are graphically represented as follows:

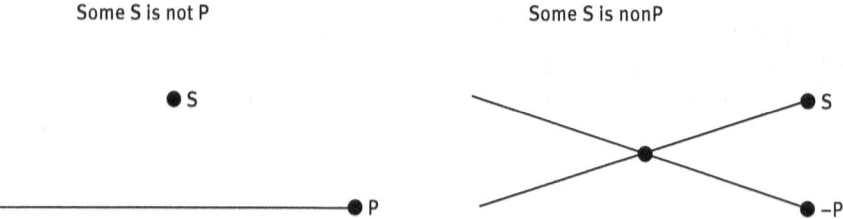

Figure 3.17: Diagrams for Predicate Denial and Term Negation

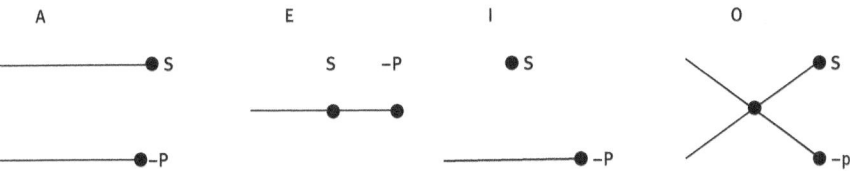

Figure 3.18: Line Diagrams for Obverted Categoricals

Obversion takes term negation seriously. Thus, any categorical statement can be diagrammed by line inclusion or line intersection since any universal negative can be construed as a universal affirmative whose predicate term happens to be negated and any particular negative can be construed as a particular affirmative whose predicate term happens to be negated.

Singular affirmations, such as 'Socrates is a philosopher', have the form '*s + P', which can be represented by a line diagram having a delineated point labelled s on a line labelled P. As well, a singular negative ('Socrates is not a Roman'/ 'Socrates is no Roman'/'Socrates is nonRoman') would correspondingly be represented with the delineated point representing Socrates on a line labelled − R:

Figure 3.19: Line Diagrams for Affirmative and Negative Singulars

Note that, given the convention that any label on any line applies to the entire line to its left, the delineated point representing the singular cannot be placed anywhere on the line other than the left terminus.

The rule of obversion is important. Yet it depends on an even more important, more fundamental principle – the *principle of noncontradiction*. Consider an individual such as Obama. He is American (and he's proven it). He is nonCanadian. Since he is nonCanadian, he is not a Canadian. Not everything that is not a Canadian is nonCanadian. The moon is neither Canadian nor nonCanadian; it *is not* a Canadian, however. The number of planets is not green, but it isn't nongreen either. Whatever is nongreen is either red or blue or white or yellow or …. Colour terms simply do not sensibly apply to numbers. Terms of citizenship do not sensibly apply to astronomical objects. Terms like 'drinks procrastination' (as well as its contraries) do not sensibly apply to the quality of quadruplicity. In general, if an individual, x, is nonP, then x is not P. But the converse doesn't follow. What does follow from all of this is that if a pair of contradictory statements cannot hold at the same time and in the same respect, then the same is true for a pair of contrary statements (statements such that one affirms a property and the other affirms a contrary of that property of the same thing). Here is Aristotle in

Metaphysica 1005b19 – 20: "[T]he same attribute cannot at the same time belong and not belong to the same subject and in the same respect." Then, at 1011b13 – 22:

> [C]ontradictory statements are not at the same time true ... it is impossible that contradictories should at the same time be true of the same thing. For of contraries, one is a privation not less that it is a contrary .. it is also impossible that contraries should belong to a subject at the same time.

All this is as close as Aristotle got to expressing in a clear, unambiguous manner, the principle of noncontradiction. Keep in mind that for Aristotle the contradictory of a statement is not what is today called its negation but rather a corresponding statement having those same terms in the same order but a different quantity and a different quality. Here is what seems clear enough. Contradictories cannot both hold at the same time. Statements of the two forms 'Every S is P/nonP' and 'Some is isn't P/nonP (where 'S' is either singular or general) are contradictories and cannot hold at the same time. But as well, pairs of statements that are of the same logical form but such that the two predicate terms are contraries cannot hold at the same time. For Aristotle, the *privative* of P is nonP. As we have seen, whatever is nonP has some property contrary to P. If nonP is the *logical contrary* of P, then that other property is one of the *nonlogical contraries* of P. Red, blue, etc. are the nonlogical contraries of green; being nongreen amounts to being one of these. So green has many contraries but only one logical contrary. Some properties have just one nonlogical contrary (even/odd); others have an infinite number of nonlogical contraries (1 meter long/1.1 meter long/...). In summary, then, contradictory pairs cannot both hold at the same time and, consequently, pairs that attribute contrary properties of the same thing cannot hold at the same time. For, a statement attributing a nonlogically contrary property entails one that attributes its corresponding logical contrary, which, in turn, entails the contradictory of the statement that attributes the original property. For example, 'Every X is red' entails 'Every X is nongreen' and 'Every X is nongreen' entails 'Every X is not green' and 'Every X is not green' is the contradictory of 'Some X is green'.

Linear diagrams adhere to the principle of noncontradiction in the following way. Given a specifiable domain, for any term line, P, there will be a (possibly tacit) term line, nonP, such that the two lines share no point in common. This guarantees that, for example a universal affirmation and its corresponding particular negative are contradictory. In order to represent both statements in the same diagram one would be required to do the impossible: make at least one pair of lines that share no point in common (say, the P and nonP lines) intersect.

According to TFL, a contradiction always has the form of a particular that is algebraically equal to zero: +S−S; a tautology always has the form of a universal that is algebraically equal to zero: −S+S. Consider, for example, an attempt to diagram simultaneously a universal affirmation and its corresponding particular negative:

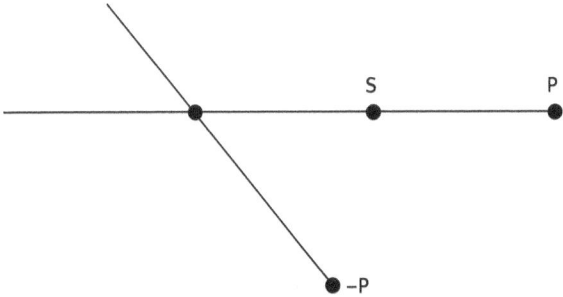

Figure 3.20: An Attempted Diagram for a Contradictory Pair

Note that one of the statements represented by this diagram is that some nonP is P (+(−P)+P), a particular algebraically equal to zero. Moreover, the principle of noncontradiction demands that no two lines representing a term and its negation have any point in common. That means in the diagram above, the term line for −P must be at once both intersecting and not intersecting the P term line. Diagrams like this are similar to Escher drawings; they give only an illusion of what is actually impossible. Such an impossible diagram would be, to use Marc Champagne's delightful neologism, a "contrapiction" (Champagne 2016). Tautological forms are quite another matter.

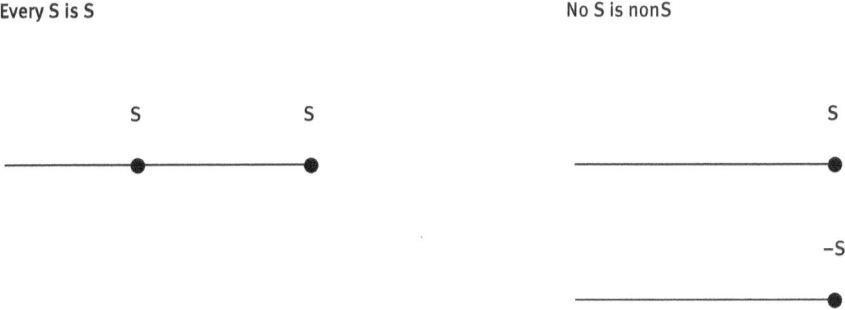

Figure 3.21: Diagrams for Tautologies

Suppose we represent each term line along with its contradictory term line. By virtue of the law of noncontradiction the two lines must not share any point. Following this practice, the *full* diagrammatic representation for an A categorical would be:

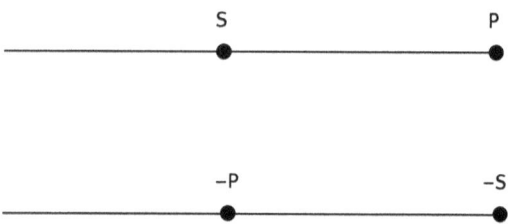

Figure 3.22: Full Diagram of A

Such a diagram represents all of the following: 'Every S is P', 'Every nonP is nonS', 'No S is nonP', 'No nonP is S', 'No nonS is P', 'No P is nonS', 'No S is nonS', 'No nonS is S', 'No P is nonP', 'No nonP is P'.

Any categorical can be given such a full representation.

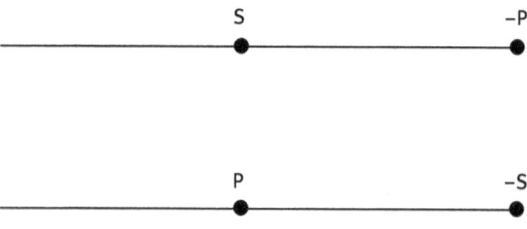

Figure 3.23: Full Diagram of E

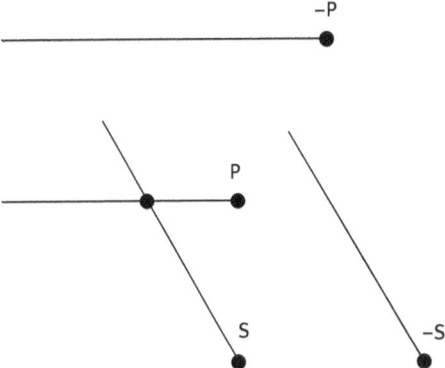

Figure 3.24: Full Diagram of I

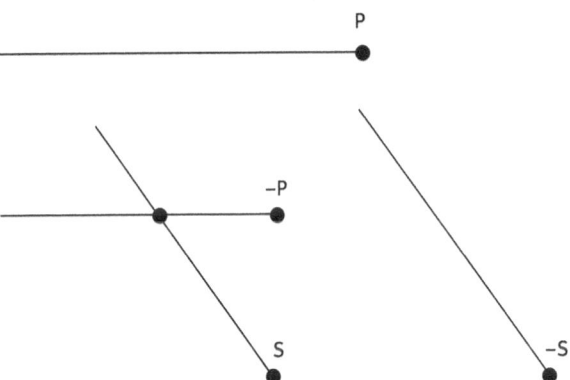

Figure 3.25: Full Diagram of O

Notice that a full diagram always consists of two pairs of non-intersecting lines. As well, just as we saw in the case of a full representation of A, full diagrams in general always represent a number of propositions, many of which are tautological or redundant. In practice, when engaged in logical reckoning diagrammatically, *simple* rather than *full* diagrams are adequate. In effect, a simple diagram for a categorical merely ignores the representations of the law of noncontradiction. Gardner called this simplifying process "minimizing" (Gardner 1982, 72). Examples of simple, minimized diagrams for A, E, I, and O categoricals are seen in Figures 3.12, 3.14, 3.15, and 3.16 above.

The traditional logical relations among categoricals, usually represented on a square of opposition (Englebretsen 2015, ch. 5, and 2016a) are preserved and graphically exhibited by ED. We have seen how contradictories (e.g., A/O and E/I) are treated. The contrariety of A and E is demonstrated by the impossibility of simultaneously diagramming both (a contrapiction).

Figure 3.26: Contrapictions of A/E Contrariety

The relation of subcontrariety requires that I and O be logically compatible. This is graphically shown by a diagram expressing both categoricals simultaneously:

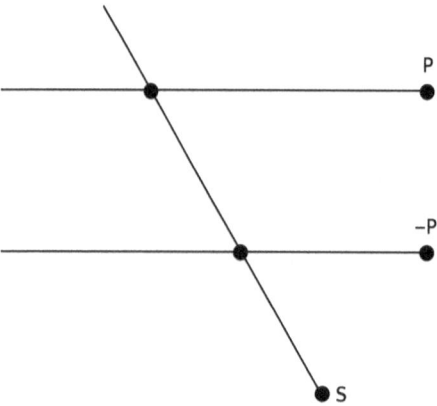

Figure 3.27: Diagram of Subcontrariety

An understanding of the logical relation of subalternation depends on how one is to understand the notion of existential commitment. The controversy stems from the question of whether Aristotle limited the syllogistic to inferences whose statements involved only nonempty terms. If the answer is positive, then all of the traditional relations illustrated on the square of opposition – in particular, subalternation – hold. If the answer is negative, then all bets are off. Modern predicate logic seems to follow the Boolean line that only existentially quantified statements express ontological commitment (in the sense that they are false whenever nothing in the universe of discourse satisfies the functions/predicates applied to the variables so quantified). From this point of view, it follows that under such circumstances of emptiness the corresponding universal statement is true. So, the modern view, generally, is that subalternation fails to hold in some cases (viz, the so-called vacuous cases). However, even if the traditional positive answer is accepted, there are still questions. Most importantly: How is the nonempty character of a term expressed? One answer is that the very use of a term ensures this. Such a view is most often attributed to Aristotle; Aristotle simply *assumed* that terms of a syllogism are never empty. Łukasiewicz even expanded the list of outlawed terms to include not only empty terms but singular and negative terms as well (Łukasiewicz 1957, 72). The idea that Aristotle's syllogistic eschews empty terms is then attributed to his medieval followers (Kneale and Kneale 1962, 59–60). Nonetheless, whether Aristotle did in fact hold such a view of empty terms, there is some reason to believe that at least some late-medieval logicians believed that the nonempty character of syllogistic term cannot simply be assumed but must be expressly asserted. Parsons has argued that this is best done by asserting of a term, T, that some thing is T, which he equates with asserting that some T is T (Parsons 2014, 65–66). According to this view, while 'Every S is S' may be a tautology, 'Some S is S' (= 'Some thing is S' = 'An S exists' = 'There is at least one S') is contingent. We saw that Pagnan adopted this position for his SYLL and it is a central part of TFL, where it is formulated as +S+S, which is neither universal nor algebraically equal to zero, thus contingent.

It seems that our choice is either to simply assume that no empty terms can be used in our inferences or to require that the nonempty character of a term being used be explicitly expressed. However, a compromise is possible. One can take terms to be nonempty but only explicitly express this when demanded by inferential circumstances. This means that subalternation can be taken to be a matter of immediate inference even though, in some circumstances, it can be taken as mediate, an enthymeme whose missing premise states that its subject term is nonempty. It can be argued that Aristotle too allowed that subalternation can be mediate.

> Now, in order to avoid the existential-import problem, one should assume that the term whose existence is not explicit does exist. Thus, if the problem relates to the term, x, one should assume that the premise 'There exists an x' is true. This has been called the "Aristotelian proviso" (Alvarez and Correia 2012, 304).

TFL recognizes *Aristotle's Proviso*. Subalternation is generally expressed as 'Every S is P, therefore some S is P' (−S+P ∴ +S+P) and, when appropriate, as

'Every S is P, (some S is S), therefore some S is P (−S+P, +S+S ∴ +S+P). This distinction between *implicit* and *explicit* expression of existential import (thus subalternation) can be readily exhibited in the graphic system of ED.

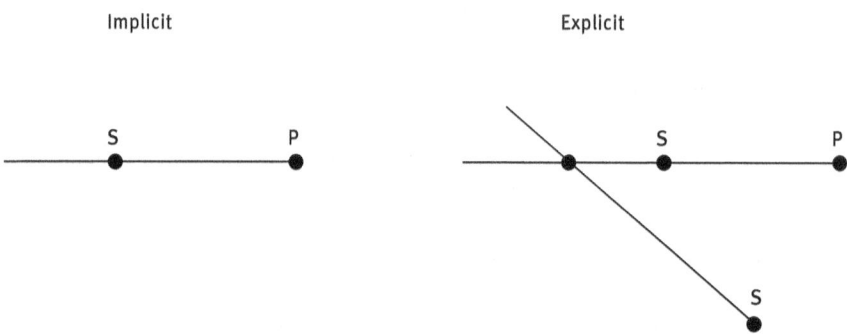

Figure 3.28: Two Versions of Subalternation

One can say that by the lights of the graphic system ED a term is to be taken as nonempty when it is represented as a term point or a term line with a specified, delineated point. *To be is to be delineated.* When existential import is taken implicitly, the very fact that a term is used (and can be diagrammed by a term line or term point) is sufficient. The labelled right terminal point of the S line shows that S and P lines share at least one point. In the case where existence is made explicit for the S term, the intersection of the two S lines explicitly shows a point (other than the right terminal point of S) that is shared by both the S and P lines. Later we will see the usefulness of having a way to make existence diagrammatically explicit.

Thus far we have seen how the graphic system ED can be used to diagram categorical statements (including singulars), exhibit existence (nonempty terms), and display the logical relations illustrated by the square of opposition. The real *raison d'etre* of any system of formal logic, whether diagrammatic or linguistic, is to provide a method to assess arguments for validity or invalidity (a decision procedure) and to provide a way to prove the former (a proof procedure). So, our next step now is to offer a decision procedure for ED. Traditional term logic worked out a fairly simple way to determine the validity/invalidity of any syllogism. The idea was that all, and only, valid syllogisms obey the "rules of syllogism":

Rules of Syllogisms
1. At least one premise must be universal.
2. At least one premise must be affirmative.

3. No term may be distributed in the conclusion if it is not distributed in the premise in which it occurs.
4. The middle term must be distributed at least once.
5. Any term distributed/undistributed in the premises must be distributed/undistributed if it occurs in the conclusion.

Notice how much the notion of term distribution is involved here. Now, as it happens, the theory of term distribution is contentious, though the various versions of the theory seem to have originated by Aristotle's remarks in Chapter 7 of *De Interpretatione*. This is partly because so many different definitions of *distributed* have been offered and partly because the notion of term distribution has so often been grounded in semantic rather than syntactic features determined by the term's role in a given sentence. It was sometimes said that a term used in a given sentence is distributed just in case it makes reference to its entire denotation. At other times it was held that the distribution of a term in a given sentence was a matter of its supposition. At still other times the claim was that a term used in a given statement is distributed in that statement just in case that statement entailed a universal statement in which that term was the subject term. Denotation, reference, supposition, even entailment in this case are semantic notions, matters of sense or meaning or interpretation, and are themselves often unclear and invite controversy (the idea of distribution aside). Still, there is general agreement that for categorical statements subject terms of universals, but not particulars, are distributed and predicate terms of negatives, but not affirmatives, are distributed. Peter Geach was no friend of any notion of term distribution. He famously wrote, "Now we need only look at the doctrine of distribution with a little care to see how incoherent it is" (Geach 1962, 4). Geach has, of course, not gone unchallenged (see especially Parsons 2006). In fact, over the years since 1962, a fairly steady stream of friends of the doctrine of distribution have come to its defense (Makinson 1969, Williamson 1971, Sommers 1971, Katz and Martinich 1976, Friedman 1978 Sommers 1982, Rearden 1984, Englebretsen 1985 g, Wilson 1987, Hodges 1998, Sommers and Englebretsen 2000, Parsons 2006, Hodges 2009, Alvarez and Correia 2012, Martin 2013).

A decision procedure does depend, in part at least, on determining whether a given term is distributed or undistributed (its distribution value) in a given statement (viz., in a premise or conclusion of an argument). As it happens, it is possible to determine the distribution value of a term in a statement without the involvement of any semantic notions. There is a purely syntactic, indeed simply mechanical, method for doing so (for more on the contrast between semantic and syntactic accounts of distribution see Martin 2013, 135–139). Sommers indicated how this mechanical procedure could be done, at least in TFL's algebraic formal language (Sommers 1982, 181):

> [I]n TFL the question whether a given term is distributed or undistributed in a proposition is the question of whether its algebraic value in that proposition is negative or not. ... In determining the distribution value (or valence) of the elements of an expression we simplify its algebraic representation by driving the minus signs in as far as possible. The result will be an algebraic expression in which each element is either negative or positive. Negative elements are distributed, positive elements are undistributed.

Much more recently a similar idea was proposed by Wilfrid Hodges (Hodges 2009, 603), who wrote, "Briefly, a term in a sentence is distributed if it occurs only negatively, and undistributed if it occurs only positively." Keeping in mind that by the lights of TFL universal quantity is formulated as a minus sign, the mechanical determination of a term's distribution value in a statement is easily summarized as follows:

> Distribution Value
> A term is *undistributed* in a statement just in case the total number of universal quantifiers and negations in whose range it occurs is even (including zero), otherwise it is *distributed*.

How are the distribution values of terms represented in the linear diagrams of ED? Such distribution values are readily observable in such diagrams. Quite generally, these values are exhibited by the following diagrammatic features:

> A term, T, is distributed in a diagram if and only if its term line contains no labelled delineated point to the left of its right terminus, otherwise it is undistributed. If a term, nonT (−T), is distributed/undistributed in a diagram, then T is undistributed/distributed in that diagram.

One can verify this for the four standard categorical forms by looking at Figures 3.12, 3.14, 3.15, and 3.16 above. We will eventually see how this is the case for any kind of statement formulated in TFL and diagrammed by ED.

Now, as it happens, the determination of the distribution value of a term is essential when it comes to *deductions* in a term logic, whether Aristotle's syllogistic, traditional syllogistic, or TFL. But the latter can make use of a second mechanical device, other than the one for determining distribution values, in deciding argument validity. While traditional logic made use of the five Rules of Syllogisms, TFL get by with just two:

    Rules of Syllogisms (for TFL)
      1. The algebraic sum of the premises (including tacit premises) must equal the conclusion.

2. The number of particular conclusions must equal the number of particular premises.

That's it. Deciding validity or invalidity is even easier once arguments are diagramed.
   Rule of Syllogisms (for ED)
   Diagramming only the premises (including tacit premises) together must exhibit the conclusion.

That means that if either the conclusion has not been revealed simply by diagramming the premises together or more than a single diagram is required for the premises (i.e., they cannot all be diagrammed together), then the argument is invalid.

So, the time has come to turn from argument validity/invalidity decision to (valid) argument deduction, *proof*, the incremental stepwise construction of justified inferences that begins with an argument's premises and culminate with its conclusion. A deductive argument consists of a collection of a finite number of premises and a conclusion. There may be statements that are taken to be among such a collection but which need play no role in the deduction; they are redundant. There may be premises that do play a role in the deduction but are not explicitly stated among the collection of explicit premises; they are tacit (suppressed, understood) premises, premises that are either tautological or else generally accepted on other grounds; these premises are *hidden*. The conclusion is a statement explicitly made or one that is missing and must be found via the process of deduction. A deductive argument is made with the accepted, but usually unexpressed, claim that if all of the premises are accepted as true the conclusion must thereby be accepted as true. The claim is not that the premises *are* all true or that the conclusion *is* true. The process of deduction can be used to establish the understood claim. It can also be used, when necessary, to discover a tacit premise (particularly one that might not be generally accepted), so-called *lost* premises. Once a decision has been made that an argument is valid, one can *prove* that it is valid. This is done in formal, but non-diagrammatic, systems by constructing a finite list of statements (called *lines*). The list consists initially of the premises followed by additional statements, each of which is *justified* by a *deduction rule* applied to one or more of the preceding lines. The final statement of the list is the conclusion, which is also justified. Since every new line in a proof must be justified, having an acceptable set of rules is essential.

When, in his *Prior Analytics*, Aristotle set out to form a system of syllogistic proof, he took the first figure syllogisms to be "perfect" in the sense that they are "complete" and therefore require nothing further to exhibit their validity immediately. In particular, this means that no additional premises are required and that no additional middle terms need to be introduced; they need no proof. He took syllogisms in the other figures to be incomplete (*Prior Analytics* 24b23–27). This means that such syllogisms require proof. Proof of an imperfect syllogism can be provided by "reducing" it to a perfect syllogism by applying certain rules to the premises in order to change them in such a way that the result is a first figure syllogism. These rules are called rules of *immediate inference*. They simply amount to the principles of conversion, obversion, subalternation, and mutation (i.e., altering the order of the premises). As it turns out, all the imperfect syllogisms are reducible to the first figure perfect syllogisms. There is a feature that is common to the perfect syllogism and is consequently shared by all syllogisms.

Traditional logicians usually cited *Prior Analytics* 24b26–30 or 25b31–35 in formulating what was called *dictum de omni et nullo* (the principle of all and none), or simply the *dictum*. The claim is that the *dictum* is the underlying principle governing all syllogistic inference, "the foundation of the whole of syllogistic theory" (Leibniz 1966, 116). A typical version of the *dictum* is 'What is predicated (affirmed/denied) of all/some of X is likewise predicated of what X is predicated of'. Sometimes it is formulated in terms of parts and wholes: 'What is predicated of any whole is predicated of any part of that whole' (see Kneale and Kneale 1962, 79). Leibniz formulated the *dictum* as a rule of substitution: "To be a predicate in a universal affirmative proposition is the same as to be capable of being substituted without loss of truth for the subject in every other affirmative proposition where that subject plays the part of predicate" (Leibniz 1966, 88). Even Boole, it seems, took his basic rule of deduction ('Equals can be substituted for equals') as a version of the *dictum*, *qua* rule of substitution (Corcoran and Wood 1980, 615–616; also Green 2009). TFL also takes the *dictum* to be a rule of substitution (see Sommers and Englebretsen 2000, 133–135). In this case, it says that the predicate term, say P, of a universal premise can be substituted for the subject term, say M, of that premise for any instance or occurrence of that term, M, in another premise whenever the two occurrences of the term, M, have different distribution values (Englebretsen 2010, 54–55 and 2012, 74–75). In effect, this simply means that the occurrences of a pair of middle terms algebraically cancel out, in which case, the sum of the premises algebraically equals the conclusion (whenever the number of particular conclusions equals the number of particular premises). The *dictum*, as Leibniz rightly saw, is the foundational rule that governs mediate inference. It governs not only classical syllogisms but, as will be seen, all kinds of inferences, including those involving singular terms and relational terms. It governs, as well, the inferences of propositional logic.

It is possible to illustrate the *dictum* for the classical syllogisms by using the system of linear diagrams, ED. Consider Barbara, Celarent, Darii, and Ferio (the four perfect first figure syllogistic forms).

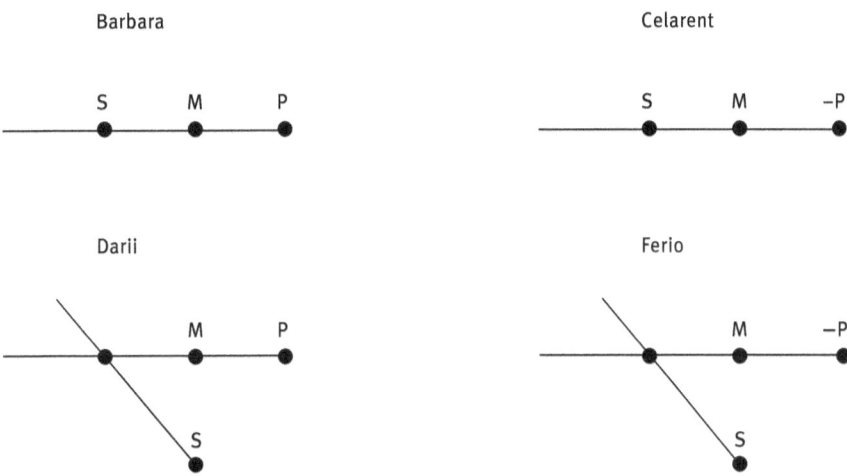

Figure 3.29: Line Diagrams for the Perfect Syllogisms

Notice that the charge (positive or negative) on the major term is immaterial. Each of the negative forms is simply a version of their positive counterparts.

Figure 3.30: Line Diagrams for Universal and Particular Syllogisms

Consider how the diagram for Barbara would have been constructed. Step 1: the major premise is depicted.

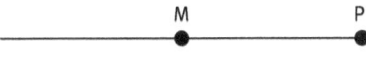

Figure 3.31: 'Every M is P'

Then the minor premise is diagrammed by incorporating it into this diagram and making use of the already represented M. The result is the full diagram for Barbara:

Figure 3.32: Barbara

Cancellation of the middle term per the *dictum* simply amounts to ignoring M, reading the conclusion directly, immediately from the diagram. Here are some examples of diagrammed imperfect syllogisms.

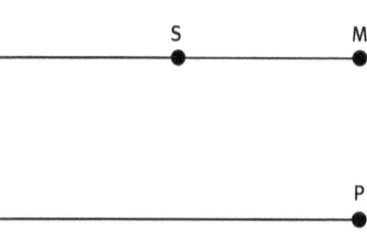

Figure 3.33: Cesare 1

Figure 3.34: Cesare 2

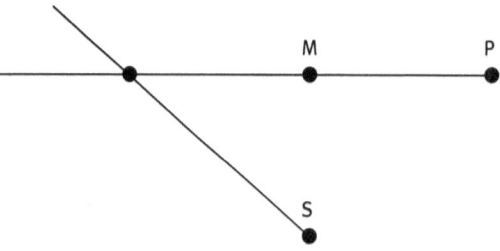

Figure 3.35: Datisi

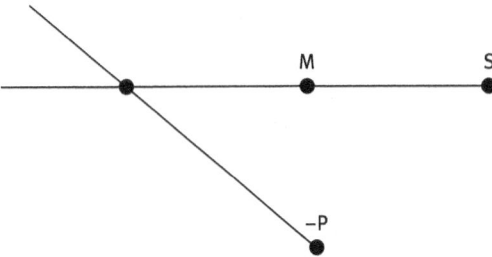

Figure 3.36: Bocardo

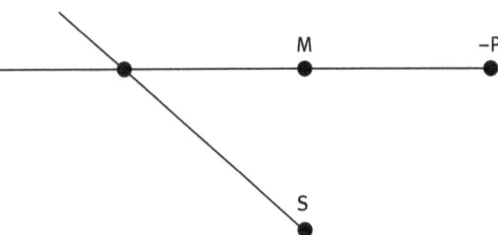

Figure 3.37: Ferison

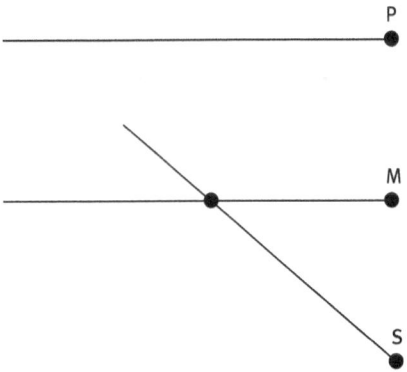

Figure 3.38: Fresison

Note that in a syllogism such as this the ED diagram indicates nothing about whether the S line extends all the way to the P line, thus indicating our lack of information. We know that at least one S is not P but we have no knowledge about any S that is P. A Venn diagram would represent the same with an x in the SM–P cell but nothing in the SMP cell.

Such line diagrams are easily constructed for every one of the twenty-four classic categorical syllogisms. Classical categorical syllogistic is thus complete. It is also sound. An exhaustive (and exhausting) check of each of the 232 classic invalid syllogistic forms shows that all are revealed to be invalid by means of this diagram system. The system is just as effective when it comes to the weakened syllogisms, valid syllogisms with a particular conclusion but no *explicit* particular premise. As with TFL, ED takes such syllogism to have a hidden premise that, when diagrammed makes the existence of the minor term explicit (as in Figure 3.27 above). For example, consider Barbarip, which has two universal premises that are the same as those of Barbara but has a particular conclusion. The present system makes the existential import of the minor term, say S, explicit by depicting the hidden premise that some S is S:

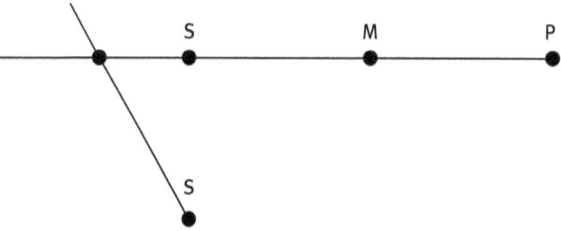

Figure 3.39: Barbarip

This system of line diagrams can be applied to arguments well beyond just classic syllogisms. One of the most crucial limitations on plane figure diagrams for logic is that they become less perspicuous and soon quite impractical as a usable tool for logic once the number of terms in an argument get much beyond four or five. No such limit applies to ED diagrams. Consider the following five-term sorites argument: Every A is B, every B is C, no C is D, some D is E; so some E is not A. Diagramming the universal premises together (in any order) and then adding the particular premise immediately reveals the conclusion:

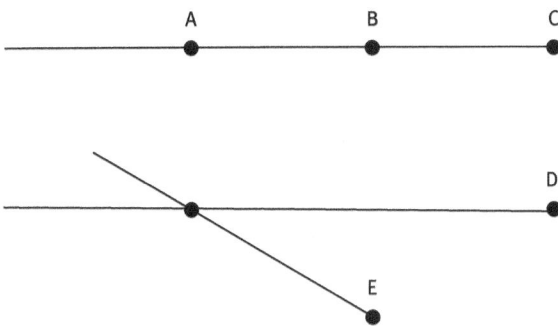

Figure 3.40: A Five-term Diagram

One task given to the logician is to determine whether or not a given set of statements is logically consistent, whether or not they can all be true together. Once such a decision is made, the next task is to deduce an explicit contradiction, usually some simple contradiction from the set. For example, any set of statements of the form: Every A is B, some A is not C, every B is C, is inconsistent. This can be shown by diagramming the statements together:

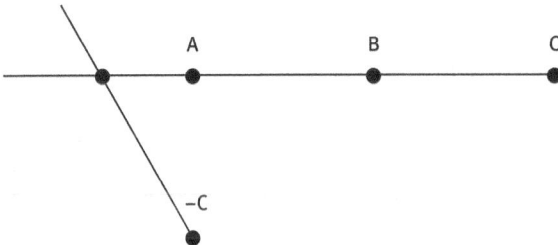

Figure 3.41: An Inconsistent Set

Note that this diagram reveals the contradictory statement that some C is nonC; it is a contrapiction. Now, since any valid argument can be proved *indirectly* by proving that the set of statements consisting of its premises plus the contradictory, or even the contrary, of its conclusion. Here is Baroco diagrammed indirectly:

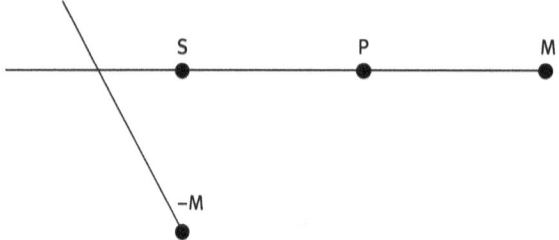

Figure 3.42: A Contrapiction of Baroco

In the next section a closer look at singular terms and how arguments in which they play a role are diagrammed.

3.4 The Point of Names

> Point exists only in line, which is in surface, which is in body, which is in matter.
>
> Avicenna

Aristotle held that the terms that were of interest to him in the building of syllogistic were, for the most part, terms of medium generality (*Prior Analytics* 43a41–42). According to some (Ross 1949, Łukasiewicz 1957, Bird 1964, Patzig 1968), for a variety of different reasons, Aristotle intended to exclude from syllogistic not only terms of highest generality (e.g., 'exists', 'substance', etc.) but also terms of lowest generality – singular terms (for counter arguments to each of their claims see Englebretsen 1980b). *Prior Analytics* is hardly bereft of examples of statements and entire syllogisms using singulars (e.g. 43a34–35, 47b24–25, 47b32–33, 67a33–37, 68b41–69a12, 70a16–29, 78b4–10). There is no prohibition of singulars from any term logic, Aristotle's syllogistic, traditional syllogistic, TFL. We have already seen how singular terms are represented (by labelled points) in ED. Moreover, one must keep in mind Leibniz's insight (Leibniz 1966, 115) that singular terms when used as subject terms accept arbitrarily either universal or particular quantity, *wild* quantity. It is because singular subject terms are wild in quantity (their quantity is arbitrary) that any sign of quantity (e.g., 'every', 'some') is ignored, suppressed in ordinary uses of a natural language. Nonetheless, from a strictly logical point of view, such singular terms *do* have a quantity. This suppression of quantity for singular subjects can be rendered explicit in ED. Consider the singular statement 'Socrates is a philosopher', which can be construed as either '(Every) Socrates is a philosopher' or '(Some) Socrates is a philosopher'. Diagrammatically:

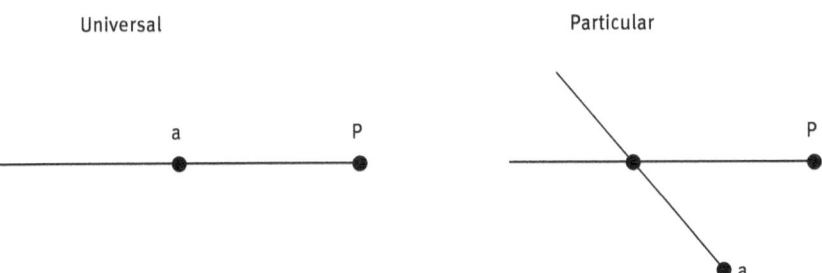

Figure 3.43: Wild Quantity

Since singular subject terms are wild in quantity, the two s-line segments in these diagrams are each reducible to the single s-point (as in Figure 3.18). Note that the distribution value of a singular subject term is likewise wild.

TFL takes all terms to be syntactically on a logical par, fit for any logical role in a statement. One consequence of this is that the semantic distinction between singular terms and general terms can be ignored (Englebretsen 1986b). Thus, singular terms, just like general terms, come in logically charged (positive/negative) pairs. Just like, for example, 'massive'/'massless', 'Kripke' has both a logically positive and a logically negative form. And, like 'married'/'unmarried', the positive charge on such terms tend to be suppressed: 'Kripke'/'nonKripke'. The obvious objection at this point is that 'nonKripke' is not a name, even a negative name, of anyone or anything. More on this soon. A second consequence of this full integration of singular terms with general terms by TFL is that, just as general terms can be predicate terms, singular terms can be predicate terms as well.

We know that Frege took the ontological distinction between objects and concepts to be absolute and inviolable. But that distinction ultimately rests on a semantic distinction, the one between singular terms (especially names) and general terms (Sommers 1982, 37–40). While a general term is understood as denoting any number of individual things, a singular term is understood as denoting a single (exactly one) thing. It is just a matter of what is taken to be the size of sets of denotata, a matter of semantics. Most logicians following in Frege's footsteps presume that this semantic distinction is equally absolute and inviolable. The distinction between singular terms and general terms is also absolute and inviolable – the so-called *Asymmetry Thesis*. The thesis is often, oddly, formulated as claim that 'subjects' and 'predicates' are logically asymmetric; yet, even then, the target expressions are singular and general *terms*. Speaking of both traditional logic and modern logic, Frank Ramsey wrote:

> Both the disputed theories make an important assumption which to my mind, has only to be questioned to be doubted. They assume a fundamental antithesis between subject and predicate, that if a proposition consists of two terms copulated, these two terms must be functioning in different ways, one as subject, the other as predicate. (Ramsey 1925, 404)

Among the many who came to the defense of the asymmetry thesis, P.F. Strawson was one of the most careful and persistent. In defending the thesis, Strawson argued against both the possibility of negating singular terms and the possibility of predicating singular terms. In the former case, he argued that while a sentence could be negated simply by negating its predicate, the attempt to negate the subject does not yield the negation of the sentence. Indeed, the result of doing so just yields nonsense. This is so because general terms "come in incompatibility groups" but singular terms do not (Strawson 1970, 102–103; 1974, 19). For example, 'red', 'blue', 'green', etc. are mutually incompatible (contrary), but what is incompatible with 'Kripke'? Who or what is nonKripke? Kripke has a very large number of properties such as being from Omaha, being the author of *Naming and Necessity*, being a male, being American born, wearing a beard. He also lacks many properties such as being Canadian born, being Belgian born, being more than six feet tall. For Strawson, whatever nonKripke would be, he, she, it lacks all the properties Kripke has and has all the properties Kripke lacks. Consequently nonKripke would, *per impossible*, be both Canadian born and Belgian born (Strawson 1970, 111n). No individual can have incompatible properties at the same time; any such purported individual (like nonKripke) is impossible.

And Strawson was right. There can be no such individual as nonKripke. But that is because *the negation of a singular term is not a singular term* (Englebretsen 1985c, 1985d, and 2015, 19–20). The negation of a singular term is a general term. In ordinary discourse we often form the negation of a singular term. In English we do this with expressions such as 'other', 'besides', 'else', 'except', 'but', etc. Given a suitable determinable universe of discourse, the denotation of a negative singular term such as nonX would be everything in the domain other than X. For example, in the sentence 'Ed but not Tom came to the party' the 'but not Tom' does not denote an impossible individual (as Strawson thought). It denotes each of the party invitees with the exception of Tom. In the sentence 'No solar planets other than Earth are inhabited' the expression 'other than Earth' (formally 'nonEarth') denotes Mercury, Venus, Mars, Jupiter, Neptune, Uranus, and Saturn. Compare the diagrams for 'Earth is inhabited' and 'No solar planets other than earth are inhabited':

Figure 3.44: Diagrams for Singular and Negative Singular Subjects

Note that the unmarked singular term in the first diagram is nonetheless implicitly charged positively (and the quantity of the subject is implicitly wild). Note as well that in the second diagram the negative version of the singular term is not singular; it functions logically as a general term (and is universally quantified here). (For more on negated singular terms see Clark 1983 and Zemach 1981 and 1985.)

Strawson's dismissal of the second result of the asymmetry thesis, the claim that singular terms cannot be predicated, cannot be predicate terms, was brief – even blunt. He wrote that this asymmetry "seem[s] to be obvious and (nearly) as fundamental as anything in philosophy can be" (Strawson 1957, 446). *Prime facie*, the sentiment seems sound. And it seems to have a very long history. In *Categories*, Aristotle distinguished, among "things that are said," between those that are "said of" (predicated of) a subject and those that are "in" a subject. There are also those that are both and those that are neither. These latter are the subjects that the others are either said of or in or both. They are individual, numerically one. They are ontologically and logically basic. They are *primary substances*. "So if the primary substances did not exist it would be impossible for any of the other things to exist" (*Categories* 2b5–6). So individuals are not said of a subject, not predicated. In the most basic sense, they are the "things there are" (*Categories* 1a20). Aristotle was less than perfectly clear about the distinction between things that *are* and things that *are said*. He included primary substance in his survey of "things that are said" even though they are things that (fundamentally) are. So how is this to be treated in a term logic?

TFL treats all terms, singular or general, count or mass, concrete or abstract, the same. Thus, singular terms are just terms, and they can be subject terms or predicate terms. We often predicate singular terms. 'The only inhabited solar planet is Earth', 'The forty-fourth U.S. president was Obama', 'That woman is Eve', 'The creator of Harry Potter is J.K. Rowling', 'Twain is Clemens' (remember in this case that TFL requires no special treatment of identity). 'No solar planets other than Earth can be logically paraphrased as 'No nonEarth is an inhabited solar planet', which is logically equivalent to 'Every inhabited solar planet is Earth', another sentence with a singular predicate term.

As has been shown, TFL has no need of a special "theory of identity." There is no necessity for a so-called "*is* of identity." Sentences taken to be expressions of identity are logically construed as predications like any other. To say that x is identical to y is simply to say that x is y and y is x. This is due in part to the wild quantity thesis that derives from Leibniz. He also formulated the principle that two terms that denote the same thing can be substituted for one another anywhere without loss of truth (Leibniz 1966, 34, 43, 52–53, 122, 131). Consider, now the example of a syllogism offered by Leibniz (Leibniz 1966, 115):

> Should we say that a singular proposition is equivalent to a particular and to a universal proposition? Yes, we should. So also when it is objected that a singular proposition is equivalent to a particular proposition, since the conclusion in the third figure must be particular, and can nevertheless be singular; 'Every writer is a man, some writer is the Apostle Peter, therefore the Apostle Peter is a man'. I reply that here also the conclusion is really particular, and it is as if we had drawn the conclusion 'Some Apostle Peter is a man'. For 'some Apostle Peter' and 'every Apostle Peter' coincide, since the term is singular.

Note the form of this syllogism: 'Every W is M, Some W is A.P, so (some) A.P is M'. Here, the singular term occurs as a predicate term in the minor premise. It also appears as a subject term in the conclusion. The syllogism is easily diagrammed:

Figure 3.45: Leibniz's Syllogism with a Singular Predicate Term

Consider next a syllogism that requires a singular predicate term (twice) and a singular subject term that occurs once universally quantified and once particularly quantified: Some writer is Twain, Clemens is Twain, so Clemens is a writer' (formally, adopting the TFL convention of using lowercase letters to label individuals: Some W is t, c is t, so c is W). In this case, the particular quantity of the major premise demand the particular quantity of the conclusion as well as the universal quantity of the minor premise. Diagrammatically:

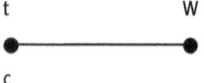

Figure 3.46: Diagrammed Syllogism with Two Singular Terms

An advantage claimed for the use of '=' as a special binary relational term is that it shares the formal features of being reflexive, symmetric, and transitive, which are definitive of the mathematicians' equal sign (also '='). Yet these features are not lost when the 'is' of identity is abandoned. Let x, y and z be singulars (thus being fit at a subject term or as a predicate term and having wild quantity when used as a subject term). Thus 'x is x' has the tautological logical form 'every x is x'. Also, the inference from 'x is y' to y is x' has the valid logical form 'some x is y, therefore some y is x'. Finally, the inference from 'x is y' and 'y is z' to 'x is z' has either of these two valid logical forms: 'every x is y and every y is z, therefore every x is z' or 'some x is y and every y is z, therefore some x is z'. In other words, all three formal features of identity are preserved without requiring identity to be a special binary relation with special rules of inference and a special symbolization. They can be diagrammed very simply as long as the theory of wild quantity is observed:

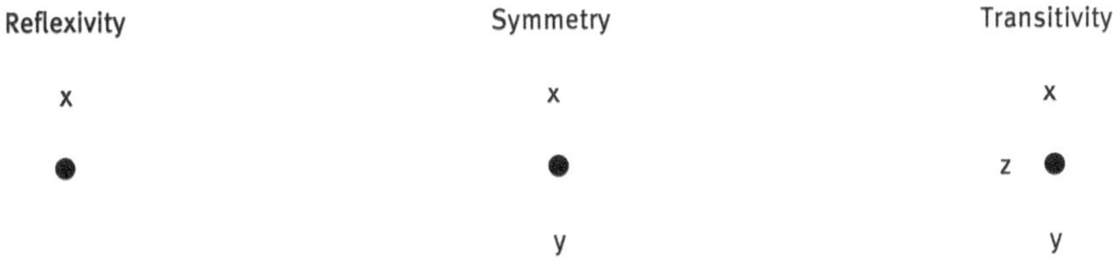

Reflexivity	Symmetry	Transitivity
x	x	x
●	●	z ●
	y	y

Figure 3.47: Reflexivity, Symmetry, and Transitivity in TFL

(For more on identity in TFL as well as the predication of singular terms see, for example, Sommers 1969 and 1982, Noah 1973, Lockwood 1975, Englebretsen 1985 f., Frederick 2013; also Moktefi 2015, 608 for the diagrammatic distinction between *is identical to* and *is*.)

Unquestionably, some singular terms are compound terms, so-called phrasal conjunctions and phrasal disjunctions. Advocates of the Asymmetry Thesis (especially Strawson) not only denied the possibility of negated singular terms and predicated singular terms but the possibility of conjoined or disjoined singular terms as well. Consider these two sentences: 'Every Beatle sings' and 'Every logician sings'. In ordinary discourse contexts, the denotation of 'Beatle' consists of just four things but the denotation of 'logician' is much, much larger. In 'Every prime number greater than 2 is odd', the denotation of 'prime number greater than 2' is infinite. Now, when the denotation of a term is known to be relatively small, it is possible to dispense with the use of that term and use in its place another term that explicitly indicates each of the things in its denotation. In place of 'Beatle', for example, one could substitute 'John, Paul, George, Ringo'. Call '₁John, Paul, George, Ringo₁' a term of *explicit denotation* since it wears its denotation on its face. In fact, this is true of any term. In principle (though rarely practical) any term, T, whose denotation consists of a, b, c,..., is replaceable by ⌊a, b, c, ...⌋. Thus, 'Every Beatle sings' can be replaced by 'Every ⌊John, Paul, George, Ringo⌋ sings'. This is clearly not an ordinary English sentence. Its ordinary version would be 'John, Paul, George, and Ringo sing'. In this case the logical quantifier 'every' is now indicated by the word 'and'. By contrast, the familiar sentence 'John, Paul, George, or Ringo plays drums' uses 'or' to indicate particular quantity. Its more formal version would be 'Some ⌊John, Paul, George, Ringo⌋ plays drums'. The use of such phrasal conjunctions and phrasal disjunctions is rare in ordinary discourse because the vast majority of terms have denotations that are too large or too indeterminate to be expressed by the use of corresponding terms of explicit denotation.

Consider, once more, 'Every Beatle sings'. It is a simple universal affirmation. It could be symbolized in TFL as: −B+S and, substituting for B its explicit denotation: − ⌊J, P, G, R⌋ +.S. But now consider 'The Beatles won the top quartet prize'. In this case, the expression 'the Beatles' could not reasonably be replaced by the explicit denotation version of 'Beatle'. In 'Every Beatle sings' the quantifier applied *distributively,* so that 'sings' can be predicated of each Beatle. That is why phrasal conjunctions are routinely replaced by sentential conjunctions in MPL ('Every (A and B) is C' becomes 'Every A is C and every B is C'). However, the expression 'the Beatles' is not a phrasal conjunction, not logically equivalent to 'John, Paul, George, and Ringo'. John didn't win the top quartet prize, nor did Paul, nor did George, nor did Ringo. Who won that prize? The quartet, consisting of John, Paul, George, and Ringo. A quartet always has four members, four individuals, *but is itself an individual.* Compare that with the sentence 'Russell and Whitehead wrote *Principia Mathematica.* Russell didn't write it. Whitehead didn't write it. The duo, the writing team whose members were Russell and Whitehead wrote it. Call terms like 'the Beatles', 'Bourbaki', 'the New York Yankees', 'the Vienna Philharmonic' *team terms.* On some occasions, 'Russell and Whitehead', 'Venus and Serena', 'John, Paul, George, and Ringo' are team terms (on other occasions they are simply phrasal conjunctions. The term 'Peter, Paul, and Mary' is the name of the trio, so it is simultaneously a team term and a term of explicit denotation. Team terms are singular terms. They denote one individual – a team (which happens to consist of individuals). Teams, being individuals, could themselves be members of teams of teams, and so forth. When a team term is quantified (plays the logical role of subject term or object term) its quantity is, of course, wild. Let a, b, and c constitute a team. The team may or may not have a name. In either case, let ⌈a, b, c⌉ denote the team, call it an *explicit team name.* For example, 'the Beatles' is a team name and ⌈John, Paul, George, Ringo⌉ is its explicit name. (For much more on terms of explicit denotation, team terms, etc., see Englebretsen 1996, 183–185, but especially Englebretsen 2015, 121–131).

Terms of explicit denotation can be treated just like any general term in TFL. In ED, they would be represented graphically just like any general term. For example, 'John, Paul, George, and Ringo sing' would be treated as 'Every ⌊John, Paul, George, Ringo⌋ sings (− ⌊J, P, G, R⌋ +S) and diagrammed as a normal A categorical. 'John, Paul, George, or Ringo plays drums' would be treated as 'Some ⌊John, Paul, George, Ringo⌋ plays drums' (+ ⌊J, P, G, R⌋ +P) and diagrammed as a normal I categorical.

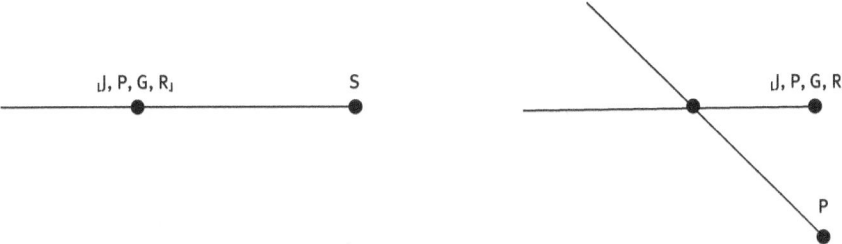

Figure 3.48: Diagrams With Terms of Explicit Denotation

A sentence such as 'The Beatles won the top quartet prize' could be construed as using a simple team name ('the Beatles') or an explicit team name ('John, Paul, George, Ringo'). In either case, they are treated as singulars (*B +W and *ʳJ, P, G, R¹+W) and diagrammed accordingly:

Figure 3.49: Diagrams With a Team Name and With an Explicit Team Name

3.5 Vectors of Relations

> It is well known that relations are more difficult to represent in a graphical system than are properties.
>
> S.-J. Shin

Modern logicians pride themselves in having at hand a system of formal logic that is adequate for the demands unmet by ancient and traditional logicians.

Prominent among such demands is that any adequate system of formal logic should be able to account for relational expressions, statements making use of such expressions, and deductions involving such statements. It is known that Aristotle, though unable to provide an adequate account of relationals, was fully aware of the importance of such a requirement, used and discussed examples of relational statements and syllogisms using such statements, and formulated rules for such deductions (see Bocheński 1951 and 1968, Sommers 1982, ch 7, Englebretsen 1982d). Traditional logicians also were mindful of the importance of accounting for relationals, with treatises on so-called *oblique* term cases. And, of course, Leibniz made an attempt to incorporate the logic of relationals into syllogistic. The problem with such efforts, according to the moderns, is that a categorical statement can have only one subject term, one quantified term, while relationals require two or more reference-making terms (terms that are either singular or quantified). Term Functor Logic illustrates just how a formal term logic can be adequate for the demand to account for the logic of relationals. Relational expressions are terms on par with all other terms. In fact, they are complex terms. Complex terms are always pairs of less complex terms that are logically copulated (by split or unsplit copulae). Moreover, the graphic system ED is able to provide a perspicuous representation of relational terms.

Consider this simple example, 'Romeo loves Juliet', symbolically: *R+(L*J). A relational term often has a "direction" (determining whether it is to be understood actively or passively). TFL indicates this by means of a system of subscribed numerals, which will be ignored for now. ED makes use of vector lines, arrows, to represent relational terms like 'loves'. Such vectors quite literally indicate direction. Thus:

Figure 3.50: Romeo loves Juliet

Not only does Romeo love Juliet, Juliet loves Romeo. So:

Figure 3.51: Juliet loves Romeo

Note that the use of vectors allows one to deduce from these two diagrams that Juliet is *loved by* Romeo and that Romeo is *loved by* Juliet. It all depends upon reading the vector from tail to head (active voice) or from head to tail (passive voice).

Consider now 'Some officer is giving a ticket to every speeder'. Its overall form is categorical. Its subject is the quantified simple general term 'some officer' and its predicate is the qualified complex term 'is giving a ticket to every speeder'. Those two terms, then, are copulated by the split copula 'some...is'. The predicate term, being complex, is itself the copulation of two less complex terms, 'giving a ticket to' and 'speeder', the copula of which is the unsplit 'every'. As well, the complex term 'giving a ticket to' is the copulation of two less complex terms, 'giving... to' and 'ticket' with the copula here being the unsplit 'a' (i.e., 'some'). It can be rendered symbolically as: +O+((G+T)−S). This could be diagrammed as an I categorical whose predicate term happens to be complex. In some logical context that would be enough. However, in most logical contexts, ones in which one of the constituents of the complex predicate term occurs without the rest, it is necessary to *analyze* such complex terms. One could analyze the complex predicate either partly or completely. In the former case, 'giving a ticket to' is left unanalyzed; in the latter case it is analyzed. Here is how our example is diagrammed with no such analysis:

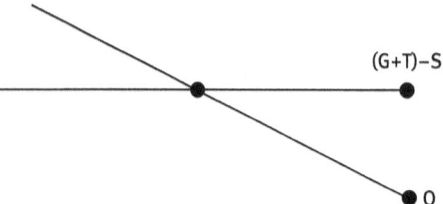

Figure 3.52: Unanalyzed Complex Relational

In other contexts, it could be necessary to at least partially analyze this complex relational:

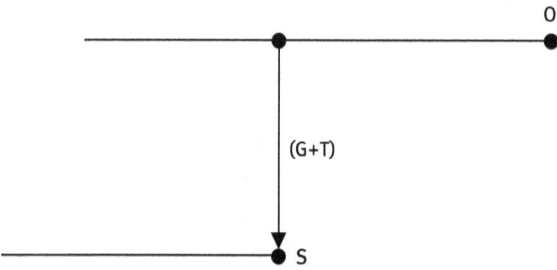

Figure 3.53: Partially Analyzed Complex Relational

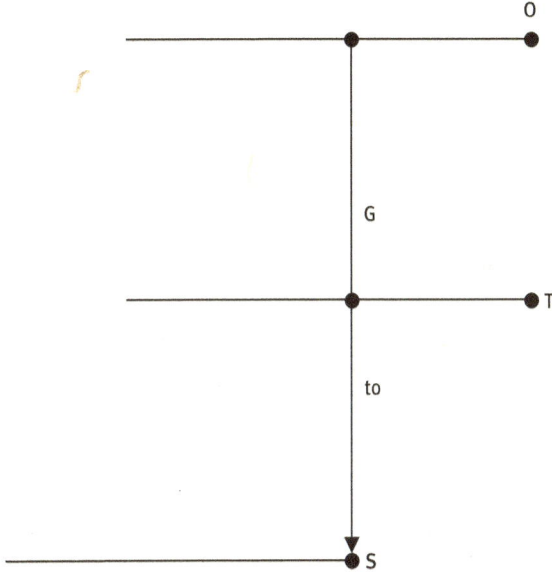

Figure 3.54: Fully Analyzed Complex Relational

It is important to notice that the locations of any points through which such vectors pass, including arrow tail points and arrow head points, indicate the quantities of the relata. The quantity is universal whenever the vector intersects the right terminus of a line and particular whenever it intersects a point to the left of the terminal point.

Of course, one could diagram such an example with the complex relational both unanalyzed and analyzed. This would be a representation of a fundamental principle: *Whatever is related (R) to some/every A is related (R) to some/every A.* Thus these two tautologies:

Figure 3.55: The Principle of Relational Analysis

So the corollary:

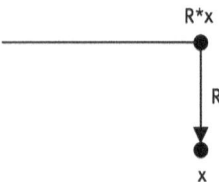

Figure 3.56: A Corollary

Consider 'Some A is R to some B'. It could be diagrammed as:

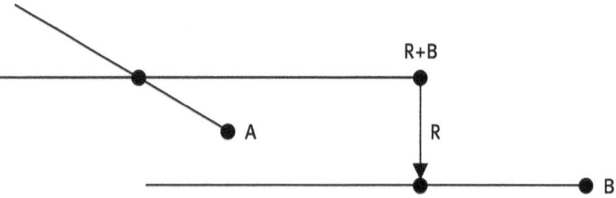

Figure 3.57: Unsimplified Relational

Since some of the information exhibited here is tautological and redundant (the inclusion of the unanalyzed relational term), the diagrammed can be simplified:

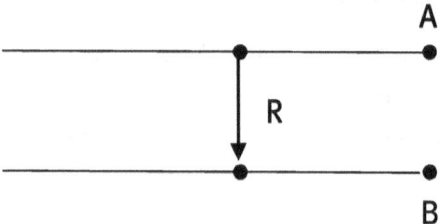

Figure 3.58: Simplified Relational

There are, however, cases requiring the inclusion of such tautological information. Consider the following argument: 'Some senator received a bribe, all bribes are illegal, whoever receives something illegal is a crook, therefore, some senator is a crook'. In this case the relational term 'received a bribe' must be represented diagrammatically as both analyzed and unanalyzed:

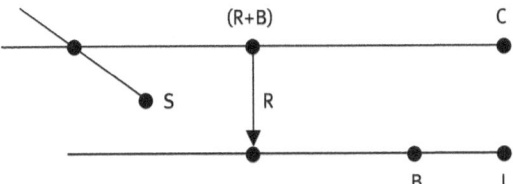

Figure 3.59: Valid Relational Inference

Such a strategy will always be required whenever a relational expression is used, as in this case, as a subject term in one occurrence and as a predicate term in another occurrence. The conclusion can be seen as already diagrammed just by diagramming the premises. But the diagram reveals much more. For example, it could also be concluded on the basis of the diagram that a bribe was received, that a bribe was received by a senator, that a bribe was received from a crook, something illegal was received, that something illegal was received from a senator, and that something illegal was received from a crook.

De Morgan's famous example of a relational inference ('Every horse is an animal, therefore, every head of a horse is a head of an animal') can provide some valuable insight into the logic of relationals. Letting C stand for 'head of', this is formalized in TFL as: −H+A ∴ −(C+H)+(H+A). TFL treats this as an enthymeme whose tacit premise is the tautological 'Every head of a horse is a head of a horse'. The full ED diagram for the inference is:

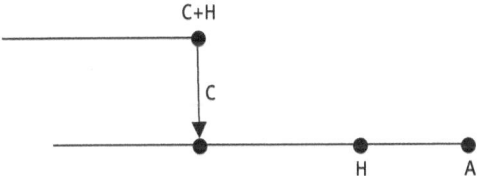

Figure 3.60: De Morgan's Inference

This example provides motivation for formulating a further fundamental principle governing the logic of relationals. Consider these four argument forms (followed by the appropriate diagrams):

1. Every X is Y, some S is R to some X, therefore, some S is R to some Y
2. Every X is Y, some S is R to every X, therefore, some S is R to some Y
3. Every X is Y, every S is R to some X, therefore, every S is R to some Y
4. Every X is Y, every S is R to every X, therefore, every S is R to some Y

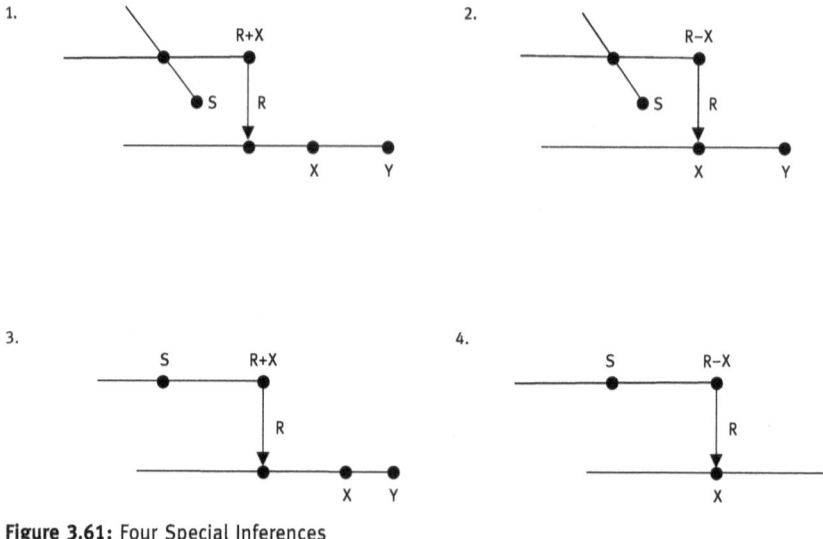

Figure 3.61: Four Special Inferences

As it happens, the conclusion of 2 and the conclusion of 4 can only be deduced by introducing a hidden premise of the form 'Some X is X', which is contingent. As the diagrams show, the only additional premise required here is a tautology found above in Figure 3.52. It is the Principle of Relational Analysis: whatever is R to some/every A is R to some/every A. Based on the four cases above, the following generalization can be formulated as an additional fundamental principle, the Principle of Relational Extension, governing the logic of relationals: *If every X is Y, then whatever is R to some/every X is R to some Y.* The diagrammatic version is:

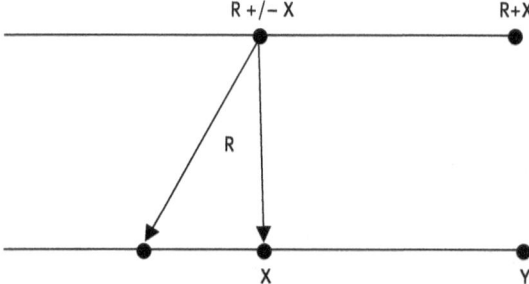

Figure 3.62: The Principle of Relational Extension

It is by virtue of the principle that a line segment representing a relational term ('R to some X' or 'R to every X') can be extended to the right so that it is a part of the line representing 'R to some Y' whenever every X is Y.

Binary relational predicates such as 'loves', 'picks up', 'next to' and the like are composed of a relational term and an object. An *object* is a quantified term. Syntactically, it is no different from any *subject*. A ternary (or more) relational predicate requires two (or more) objects. One such object will be a *direct* object while the other objects are *indirect* objects. In the example 'Some officer gave a ticket to every speeder', the relational predicate consists of the relational term ('gave...to'), an indirect object ('a ticket'), and a direct object ('every speeder'). Sometimes, *where* a particularly quantified term, whether a subject or an object, is used in a statement can determine how it is to be understood. A common illustration of this is provided by the (well-worn) example, 'Every boy loves some girl'. Here, the object ('some girl') is ambiguous. The sentence can be understood as saying of every boy (i) that he loves some girl or other, or (ii) that he loves some specific, certain girl. The two alternatives are given distinct logical formulations: (i) 'Every boy loves some girl (or other)', (ii) 'Some specific girl is loved by every boy'. The 'some girl' is taken to be *non-specific* in (i) but *specific* in (ii). Modern predicate logic resolves this by the order of quantifiers: (i) is a universally quantified existential, (ii) is an existentially quantified universal. TFL achieves the same resolution in terms of the order of positive and negative terms (see Sommers 1976b, 23 and Sommers 1982, 141–142 for brief but clear accounts). The two sentences would be diagrammed differently:

Figure 3.63: Specific vs Non-Specific Reference

The object in (ii) is specific; it refers to a specific girl, an individual. We might not know her name, but we can still refer to her. We might make use of a pronoun ('she'), so that we could paraphrase (ii) as 'Some girl is such that she is loved by every boy', and 'she' here is a singular term. Thus, when used as either a subject or an object it has wild quantity (*G+(L−B). That's why the diagram for (ii) represents her as merely a left endpoint of a line. More about pronouns later, but first, one more fundamental principle of the logic of relationals.

If Romeo loves Juliet, then Romeo loves (and as well, Juliet is loved). In natural language, relational terms have one or more objects. If Morgan offered a ride to Russell, then it follows that Morgan offered a ride. If he offered a ride to a friend, then he still offered a ride. In our ordinary use of our natural native language, we take relational terms and their non-relational partners on a par. Thus, for example, the 'loves' in 'Romeo loves Juliet' is the same 'loves' in 'Romeo loves' (likewise for 'lover of' in 'Romeo is a lover of Juliet' and 'Romeo is a lover'). We readily take a relational statement to entail the same statement without one or more of its direct or indirect objects. Thus, from 'Morgan offered a ride to a friend' one could deduce each of the following: 'Morgan offered a ride' and 'Morgan offered'. TFL takes this at face value, formulating the premise as: *M +((O+R)+F) and the conclusions as: *M+(O+R) and *M+O. Here, the term 'offered' can have any *adicity* (triadic, dyadic, monadic) depending on its sentential context. And that's natural. By contrast, MPL takes the adicity of any term as fixed, invariable, so that, for example, 'Romeo loves Juliet' and 'Romeo loves' are formulated as 'Lrj' and 'Lr', but the 'L' is ambiguous. It is a 2-place, binary relational term in the first, but it is a 1-place, monadic, non-relational predicate in the second. The inference from 'Romeo loves Juliet' to 'Romeo loves' ('Romeo is a lover', 'Romeo does love', etc.) is immediate for TFL and it is naturally so. By requiring that 'loves' be permanently binary, MPL can only deduce 'Romeo loves something' ('Romeo is a lover of something', 'Romeo does love something', etc.). At any rate, there is a general principle here that allows for the dropping of one or more objects from a relational: *Whatever is related R to every/some $A_1 ... A_n$ is R to every/some A_{n-1}.* Diagrammatically then (using dotted line extensions to indicate quantity options):

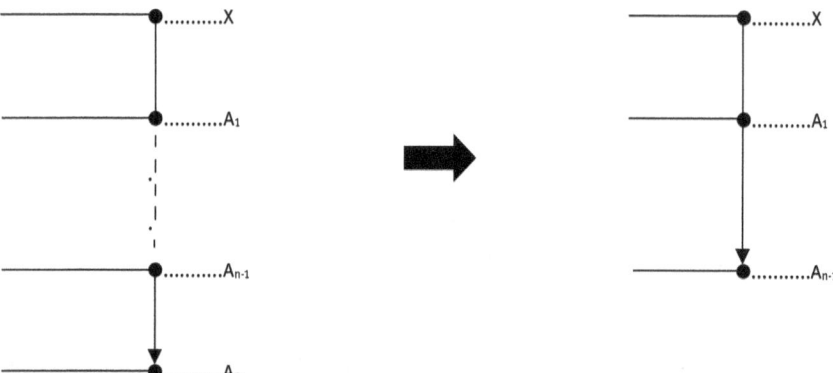

Figure 3.64: The Principle of Relational Reduction

Needless to say, other principles could be formulated to account for how converses of relationals are formed to allow for rearranging the order in which direct and indirect objects occur. For example, in English there are passive/active voice pairs such as 'loves'/'loved by' and 'gives ... to'/'gives to ...'/'given ...by'. Nevertheless, we turn now to the task of incorporating into ED diagrams strategies for representing statements involving reflexive pronouns.

In English, reflexive pronouns are usually words such as 'itself', 'herself', 'himself', 'themselves', etc. They are always object terms. Compare 'Trump loves power' and 'Trump loves himself'. In each case, 'Trump' is the subject term. In the first case, the object of the relational is 'power', but in the second case, the reflexive pronoun 'himself' has 'Trump' as its referential antecedent. This allows the pronoun to be replaced by its antecedent, which yields 'Trump loves Trump'. We know how to diagram 'Trump loves power', but what of 'Trump loves Trump'? A first attempt might produce:

Figure 3.65: Attempted Singular Reflexive Diagram

The problem with this attempt is that it treats 'Trump' (T) as two individuals, one the subject of the relation and the other the object of the relation. 'Trump' denotes a single individual, and that individual is both the subject and object of the relation. So the diagram must represent that:

Figure 3.66: Singular Reflexive Diagram

Consider a sentence in which the subject/object of the relation is a general term: 'Some senator nominated some senator'. One might attempt to diagram a sentence such as this in the same way the diagram for 'Trump loves himself/ Trump' was first attempted:

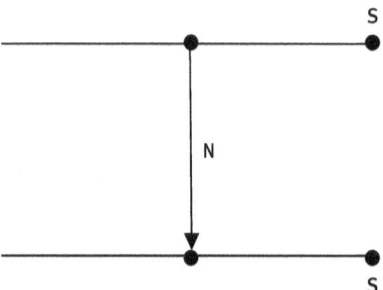

Figure 3.67: Attempted General Diagram

However, as in the first attempt, the denotations of the subject and the object are not distinct. There can only be one S line; otherwise a contradiction is depicted. The sentence can be diagrammed properly once this is recognized. Thus:

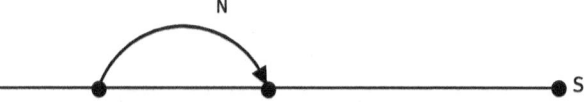

Figure 3.68: Proper Diagram

Here is an example of a valid argument involving a reflexive pronoun: 'Some senator nominated herself. Every self-nominator is a fool. All fools deserve ridicule. Therefore, some senator deserves ridicule." It can be diagrammed as follows:

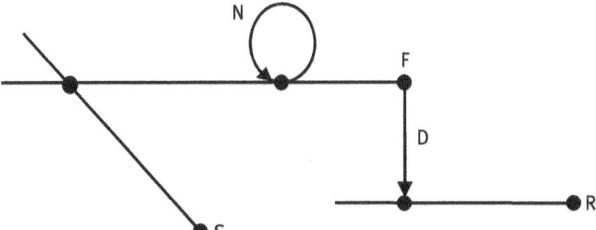

Figure 3.69: Argument With a Reflexive

As it happens, reflexive pronouns pose little challenge to the project of line diagrams. What now of (non-reflexive) personal pronouns in general? Again, we look to TFL for insight.

> It is after all a matter of historical fact that [traditional term logic] had no systematic treatment of pronouns and indeed that such a treatment is a justly celebrated accomplishment of MPL (wherein pronouns are represented as bound variables) ... one may well wonder how this false charge has received such widespread and enduring acceptance by responsible logicians. That they were anxious to persuade students that the older logic was superseded is not by itself a sufficient explanation. A minimal attention to the methods of proof available to pre-Fregean logic would have given them pause. One doubts it could happen that two generations of modern geometers could falsely claim that Euclidean geometry could not prove certain theorems that are easily provable in some non-Euclidean systems. But then mathematicians who do modern geometry are not as tendentious as philosophers who do modern logic. (Sommers 1982, 146–147)

A brief note regarding some semantic matters is in order before continuing. A used term, one used relative to a given universe of discourse (a *domain*), *denotes* every thing in the domain of which it is true. If a speaker says, relative to the neighborhood in which he or she lives, 'Every dog is barking', then' dog' denotes, in this case, every dog in the neighborhood. A *quantified* term, used relative to a given domain, *refers* to either every or at least one thing in the domain that the term *per se* denotes. So, a quantified term (a logical subject or object) has a reference determined in part by the denotation of its term and its quantifier. Thus, the term 'dog' denotes every dog in the domain relative to which it is used; 'every dog' refers to every dog in the domain; 'some dog' refers to at least one dog in the domain. The late medieval logicians seem to have been relatively clear about this distinction. Whenever a used quantified term refers to just what it denotes it is said to be distributed; otherwise it is undistributed. Briefly: terms denote; quantified terms refer.

We have seen above examples of terms that are relational ('loves some girl' (L+G)). Relational terms are complex terms. However, some complex terms are not relational. For example, compound terms such as 'rich and famous' (R+F) are complex. Consider now the following pair: 'Some dog is barking' (+D+B) and 'It is annoying' (?+A). What is annoying? Clearly, the dog. Which dog? The dog that is barking. And that is just what the pronoun 'it' denotes. The pronoun's antecedent is 'some dog' in the first sentence. While it might be thought that 'it' refers to just what its antecedent refers to, the fact is that the pronoun denotes everything to which its antecedent refers (not just some dog, but rather, *the specific dog under consideration* – the one that is barking). Such pronouns are always implicitly quantified; they are logical subjects or objects. As such, they have a denotation and also refer. Again, reference is determined by both denotation and the relevant quantifier. In effect, such a pronoun has universal quantity in addition to the quantity of the antecedent subject. In effect, the pronoun for a definite subject 'some x' will always be an expression with 'wild' quantity. Pronouns that cross-refer to 'every x' are another matter" (Sommers 1976b, 26). TFL represents this tie that connects such pronouns to their antecedents by attaching to the antecedent term a superscripted letter that is then used as the pronoun. A pronoun inherits the quantity applied to its antecedent and also has implicitly its own universal quantity (since it is meant to denote all of what its antecedent refers to). In this case, that means it has wild quantity. Thus: (+D^i+B) and (*i+A). Of course, the first sentence alone, or in a different context, need not have any subsequent pronoun. It would then be treated as a simple particular affirmation whose subject is understood to be non-specific. But, once pronominalized, such a subject must be understood as making specific reference. In that case, the first sentence above could be diagrammed in two equivalent ways:

Figure 3.70: A Simple Pronominalization

Given the pronominal ties between the two original statements, they can now be diagrammed together as:

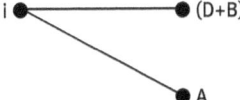

Figure 3.71: A Full Pronominalization

In MPL, pronouns, in the guise of bound variables, are ubiquitous. In natural languages, they seem to be used more sparingly. The antecedents of bound variable pronouns are always the quantifier expressions that bind them. Natural language personal pronouns usually follow their antecedents and are called 'anaphoric pronouns'. Sometimes, however, they precede their "antecedents" and are then called 'cataphoric pronouns'. Consider next an example that involves an interlocking pair of such pronominalizations: 'A boy who loved her kissed a girl who slapped him' $(+(B^h+(L*S))+(K*(G^s+(S+H))))$. Note the relations that hold between that boy and that girl. Among the boys and girls (in the domain), at least one of the boys kissed one of the girls, that boy (he, 'h') loved that girl, that girl (she, 's') slapped that boy. Thus, the boy who loved the girl kissed the girl who slapped him. This might be diagrammed in three stages:

Step 1 (Some boy loves some girl)

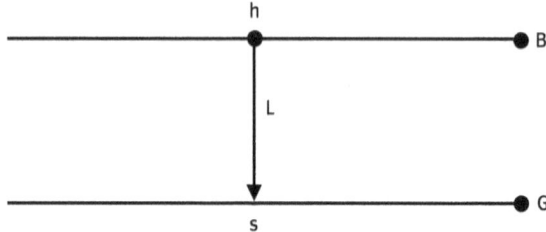

Step 2 (He kissed her)

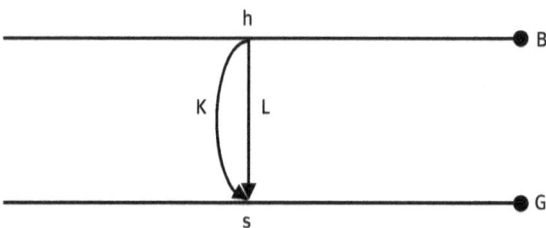

Step 3 (She slapped him)

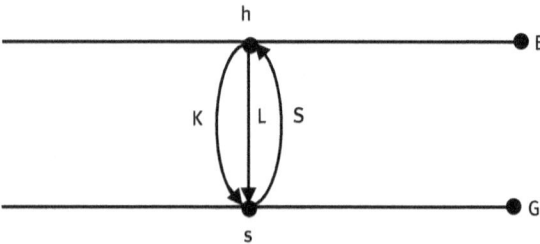

Figure 3.72: An Example of Interlocking Pronominalizations

Finally, here is an example of a diagrammed inference involving a pair of pronouns: 'A cat is screeching; I own it; it annoys me; but all cats are lovable; so, something lovable annoys me' $(+C^f+S)$, $(*i+(O*f))$, $(*f+(A*i),(-C+L)$ $\therefore$ $+L +(A*i)$.

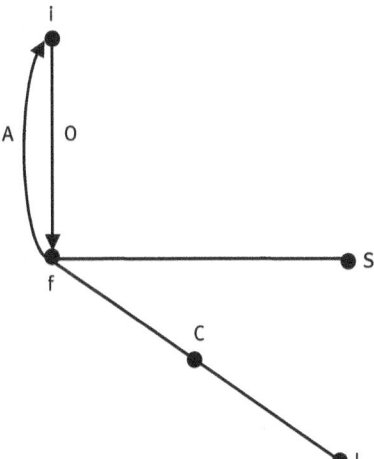

Figure 3.73: A Diagrammed Inference with Pronouns

We have claimed that all terms are positively or negatively charged. Consequently, the same is true of relational terms. In the absence of any contextual clues, the sentence 'Some critics don't like every Shakespeare play' is ambiguous between something like (i) 'Some critics fail to like every Shakespeare play'; perhaps they like the comedies but not the tragedies, or they like all of them except for Othello, and (ii) 'Some critics dislike every Shakespeare play'. This is shown in their TFL formulations: +C−(L−S) and +C+(−L−S). Note that the relational term ('likes') is positively charged in the first version (though it is part of the negative complex predicate term ('fails to like every Shakespeare play'). By contrast, the relational term in the second version is itself negative ('dislikes'), (while the complex predicate term of which it is a part is positive). Diagramming the two alternatives graphically represents their logical differences:

(i)

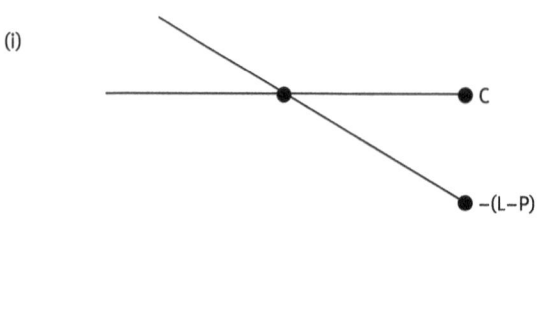

(ii)

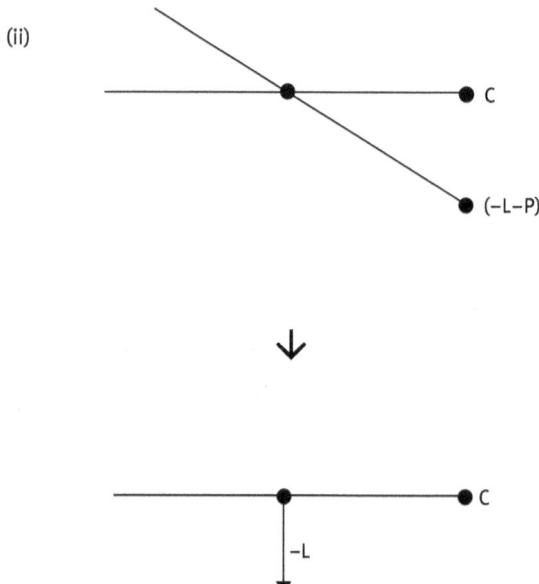

Figure 3.74: Negative Relationals

Note that the relational predicate term in the diagram of (ii) has then been analyzed. Note as well that, given the innocent tacit premise that there are Shakespeare plays, (ii) entails (i).

The real import of negative relationals is seen in cases of inferences in which they play a logically effective role. An example of such an inference is the following: Some critics dislike every Shakespeare play. *Othello* is a play by Shakespeare. All actors like *Othello*. So, no actor is a critic. It can be diagrammed thus:

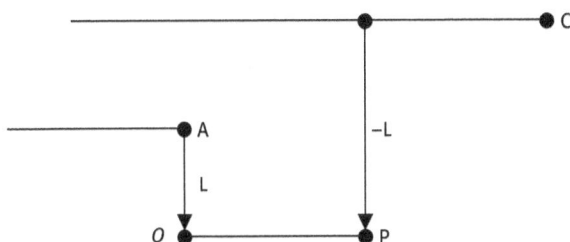

Figure 3.75: Inference with Negative and Positive Relationals

Note that in this case the A line cannot be raised to the level of the C line (which is why no A is C) since the L and –L arrows could never meet. Were the A line to be raised the result would express the contradictory proposition that all actors both like and dislike *Othello* – a contrapiction.

3.6 World Lines

> Between the [calculus of classes and the calculus of propositions] there is ... a certain parallelism, which arises as follows: In any symbolic expression, the letters may be interpreted as classes or as propositions, and the relation of inclusion in the one case may be replaced by that of formal implication in the other.
>
> Russell
>
> It was Sommers' insight that propositional logic can be developed within syllogistic.
>
> John Bacon

Whether expressing it symbolically or representing it graphically, determining the logical form of a statement is always a matter of choosing the appropriate level of analysis (see, for example, Corcoran 1999a and 2008). To see this, consider an inference that involves a syntactically complex term (the relational term 'admires some logicians') requiring no further analysis of it into its non-complex constituents: 'Every philosopher admires some logicians, whoever admires any logician is misguided; so, every philosopher is misguided'. One could, of course, diagram this with the usual analysis of the relational, using an arrow for 'admires' with a tail on the right terminus of the line for 'philosopher' and a head on a non-terminal point on the line for 'logician'. The 'philosopher' line would then be extended to the right to accommodate the second premise. However, since the constituent terms of the relational term do not play any logical role in the argument, analysis of the relational term is unnecessary. It is simply the middle term of a simple Barbara syllogism and can be diagrammed as such.

Needless to say, there can be cases of inferences in which more than one complex term might be left unanalyzed. As well, there are many cases in which no such terms can be safely unanalyzed. Entire sentences are terms (sentential terms). Even with these kinds of terms, analysis into their constituent unanalyzable terms is not always required. After all, any sentence, of any complexity, can be treated as completely unanalyzed. Consider this argument: 'If every actor is a thespian then some actors are not comics, all actors are thespians; so, some actor is not a comics'. One could formalize this recognizing the categorical forms of its constituents but such an analysis is unnecessary. It is in fact just a simple modus ponens argument.

Peirce's Alpha graphs precede Beta graphs. Nonetheless, Dipert (Dipert 1981) showed that Peirce took the logic of propositions and the logic of terms to be isomorphic, equally primary, neither more basic than the other. Kant had claimed that there is an absolute difference between categoricals and compound statements, thus, between a logic of terms and a logic of propositions. Leibniz took the logic of terms to be primary. Boole's view on the matter of primacy foreshadowed Peirce's. He saw his "secondary" logic of propositions and his "primary" logic of terms to be nothing more than two of the interpretations of his algebra. Post-Fregean logicians rejected (mostly ignored) all of these views (isomorphism, exclusivity, and term priority).

Frege, and then Wittgenstein, famously claimed that a word has a meaning only in the context of a sentence (the "Context Principle"). Were this so, then the logician would be required, in the first instance, to provide a logic of sentences before going on to a logic of terms. So-called propositional logic, or sentential logic, or statement logic is a formal system for accounting for logical reckoning involving only unanalyzed sentences. In MPL, such a system is universally presumed to be *primary logic*. Nevertheless, it can be, and has been, argued that a logic of terms is prior to a logic of sentences, that a logic of terms is primary logic. In fact, the logic of unanalyzed sentences (sentential terms) is only a part of the logic of terms. The fullest expression of these claims was made by Sommers in his "The World, the Facts, and Primary Logic," where he argued that "propositional logic is a special branch of term logic" (Sommers 1993, 181). "The doctrine that term logic is primary logic has ancient roots. Historically, term logic came first, having been discovered and developed by Aristotle; propositional logic, primarily a Stoic innovation, came later" (Sommers 1993, 172). Much of this argument for the primacy of term logic was foreshadowed by some of Leibniz's insights. The most important of these was his recognition that propositions (viz., unanalyzed sentences) could be treated as terms. He said this many times and in many ways:

> However, the categorical proposition is the basis of the rest, and modal, hypothetical, disjunctive and all other propositions presuppose it. (Leibniz 1966, 17)
>
> ... any proposition can be conceived as a term. (Leibniz 1966, 71)
>
> The proposition itself can be conceived as a term. (Leibniz 1966, 86).

He also argued that the notion of *containment*, which plays a key role in logic, applies to the relations among propositions in just the same way as it applies to relations among terms:

> ... whatever is said of a term which contains a term can also be said of a proposition from which another proposition follows. (Leibniz 1966, 85)
>
> ... that a proposition follows from a proposition is simply that a consequent is contained in an antecedent, as a term in a term. By this method we reduce inferences to propositions, and propositions to terms. (Leibniz 1966, 87).

Leibniz's aim in doing this was to produce a *unified* system of logic, one that did not treat the logic of terms and the logic of sentences differently. Once more, he wrote:

> If, as I hope, I can conceive all propositions as terms, and hypotheticals as categoricals, and if I can treat all propositions universally, this promises a wonderful ease in my symbolism and analysis of concepts, and will be a discovery of the greatest importance. (Leibniz 1966, 66).

And, as well, there is Leibniz's insight that singular subjects are arbitrarily universal or particular in quantity, the "wild" quantity thesis (Leibniz 1966, 115).

Sommers incorporated these Leibnizian insights, and much more, into his argument for construing the logic of propositions as a special branch of the logic of terms. In doing so, he went much farther than Leibniz. Leibniz had merely expressed a hope that he could achieve this kind of logical unity by treating sentences as terms, but he could not provide an adequate strategy for doing so. Sommers could. As we have seen, TFL treats sentences as nothing more than complex terms (pairs of logically copulated less complex terms). This idea reflects the obvious fact that conjunctive sentences and particular affirmations share the same formal features of being symmetric and associative (but not transitive), while conditional sentences and universal affirmations share the same formal features of being reflexive and transitive (but not symmetric and associative). So, if sentences are just complex terms, then the semantic relations that characterize terms must, per force, characterize sentences. As discussed earlier in this chapter (section 2), terms used relative to some specifiable domain of discourse, generally, stand in the three semantic relations of expressing concepts (senses), signifying properties, and denoting objects, Thus, for example, with respect to the actual world today, 'foolish' expresses the concept of foolishness, signifies the property of being foolish, and denotes whatever has that property. If nothing has that property, then the term is vacuous (has neither denotation nor signification). When a statement is made, it is made by means of a sentence used relative to a specifiable domain of discourse. The concept that a statement expresses (its sense) is construed as a *proposition* (in a strict sense of that word). What of the signification and denotation of such a sentence?

Just as used non-vacuous terms signify properties, used non-vacuous statements signify properties. In this case, the properties signified are properties of what are denoted. While terms denote objects (in the domain), statements denote the domain itself (Sommers calls a domain denoted by a statement a *world*. Indeed, worlds, domains, universes of discourse "constitute the ultimate subject-matter of the discussion" (Corcoran 1999b, 941, see also Corcoran 2004, 495 ff and Hodges 2009, 599). They are what a statement maker claims truth for. Consider the statement, 'Some politicians are foolish' asserted with respect to the world today. The condition for its truth is that there be at least one foolish politician. Suppose (suspend any contrary belief for now) that Trump is a foolish politician. On such a supposition, Trump has the properties of being a politician and being foolish. But as Hume and Kant have taught, *existing* is not an additional property of Trump. Frege (Frege 1950, section 53) had argued that existence was not a property of *objects* but rather a (secondary) property of *concepts*. For example, on such a view, to say that there are horses is to say that the concept of horse is not empty (applies to some object); to say that there are no unicorns is to say that the concept of unicorn is empty (applies to no object). Variations of the view were then offered by Russell (Russell 1918, Lecture V) and then Quine (Quine 1953, 13). In spite of such credentials, the view is rejected by Sommers in favour of the idea that to say of any object that it exists (does not exist) is just to say that it is (is not) a constituent of the domain of discourse at hand. "To be in a domain is to exist" (Sommers 1993, 175). To say that Trump exists (with respect to the world today) is to say that the world has Trump as one of its constituents. To say that there are horses is to say that horses are constituents of, present in, the world; to say that there are no unicorns is to say that unicorns are not constituents of, are absent from, the world. Thus existence and non-existence are always matters of such presence or absence (Sommers 1993, 174). The statement above about Trump *says* something about him (that he is a foolish politician); but it also *claims* something about the world (that it is characterized by the presence of Trump who is a foolish politician) (Sommers 1993, 179). While properties like being foolish or a politician are characteristics of objects, properties like having or lacking, presence or absence, are characteristics of domains/worlds. They are *constitutive characteristics*. Given a domain and a constitutive characteristic, the domain either does or does not have it. A constitutive characteristic (whether positive or negative) of a domain is a *fact*. A true statement expresses a proposition, signifies a fact (a constitutive characteristic), and denotes what has that fact. And, of course, what has that fact is the domain. A false statement fails to signify a fact and thus fails to denote the domain. It must be noted that a fact is not a constituent *in* the domain; it is a property, a constitutive characteristic *of* the domain. (For more on this account of truth see Englebretsen 2006).

Unanalyzed sentences can be construed as categoricals by taking them, then, as denoting worlds. If 'p and q' is construed as having the form 'some p is q', then the sentential variables here can be read as abbreviations for 'p-world' and 'q-world' (thus, 'some p-world is a q-world'). In the same way, 'if p then q' can be read as 'every p-world is a q-world'). Their symbolic expressions in TFL are '+p+q' and '−p+q'. A so-called atomic sentence, 'p' is simply read as 'some world is p' (+p); its contradictory negation would be 'some world is non-p' (−p). However, at this point, the doctrine that the logic of sentences is merely a special branch of the logic of terms (resulting in the unified system of formal logic envisaged by Leibniz and Sommers) faces what appears to be a serious challenge. Treating the conjunctions and conditionals of propositional logic as logically categoricals (I and A, respectively) reveals a pair of disanalogies (Sommers 1993, 172–173 and 179):

> 1st Disanalogy: 'p and q' ('+p+q') entails 'if p then q' ('−p+q'), but 'Some A is B' does not entail 'every A is B' ('+A+B' does not entail '−A+B').
>
> 2nd Disanalogy: 'p and q' (+p+q) is incompatible with 'p and not q' ('+p−q'), but 'Some A is B' ('+A+B') is compatible with 'Every A is B' ('−A+B').

It turns out that the resolution of these apparent challenges is due to the recognition of two important features of the Leibniz-Sommers logical program (Sommers 1993, 179–180). First, sentences used relative to a specifiable domain are such that, if they denote at all, they denote that domain. A domain, or world, is a totality of its constituents. While the sub-sentential terms that make up the sentence might denote things in that world, the sentence itself denotes just one thing, the world (not a thing in the world). Consequently, sentential terms like 'p', 'q', 'p-world', 'q-world', etc., are singular terms whose denotation is unique. This means that 'p and q' can just as well be read as 'the world is both p and q', 'if p then q' can be read as 'the world is not both p and not q', 'p' can be read as 'the world is p' and 'not p' can be read as 'the world is not p'. Singular statements (those having singular subject terms) can be taken to have wild quantity. Thus, conjunctions, conditionals (and any other truth-functional forms of statement) can be taken to have wild quantity. It is the wild quantity thesis that accounts for the two disanalogies, for it is the same thesis that resolves these two other disanalogies (let A denote artists, D denote Dutch residents, and V denote Vermeer:

> 3rd Disanalogy: 'Some A is D' ('+A+D') does not entail 'Every A is D' ('−A+D'), but 'Some V is D' ('+V+D') does entail 'Every V is D' ('−V+D'). Both of which are then formulated simply as '*V+D'.
>
> 4th Disanalogy: 'Some A is D' ('+A+D') is compatible with 'Some A is not D' ('+A−D'), but 'Some V is D' ('+V+D') is incompatible with 'Some V is not D' ('+V−D'), since '*V+D' and '*V−D' are incompatible.

In other words, the first disanalogy is simply an instance of the third and the second is an instance of the fourth, and the third and fourth are innocuous in light of the singularity (unique denotation) of sentential terms and the wild quantity thesis.

It is because sentential terms are singular that they are given wild quantity whenever they are quantified. "[A]ll propositional terms are uniquely denoting terms and all propositional statements are singular statements" (Sommers 1993, 179). Any sentence used to make a statement denotes the domain relative to which it is used. When an analyzed sentence (one whose constituent non-sentential elements are explicit) is used to make a statement, the sentence is simultaneously used (1) to *say* something about things in the domain (world) relative to which it is used and (2) to *claim* that what is being said (the proposition expressed) is true. When an unanalyzed sentence is used to make a statement, what it is used to *say* (assert, state) just is what it is used to *claim*. In other words, what such a statement states is not something about what is or is not in the domain – it is about the domain itself.

> The distinction between saying and claiming is idle in propositional logic since the statements we are there concerned with are represented by statement letters that give no clue as to the internal contents of the statements represented (nor is any needed for the purpose at hand). All statements of statement logic are understood as being about the world. Given 'p' we interpret it as *asserting its truth claim*, viz., that the world is a p-world. Given 'p&q' we interpret it to say that the world is both a p-world and a q-world and so on for other compound forms. (Sommers 1993, 179)
>
> Consider 'p&q'; in its categorical form 'some p-world is a q-world'. That this makes a claim about the one world is evident from its equivalence to '*The* world is both a p-world and a q-

world'. ... All statements of statement logic are understood as being about the world. ... all sentential terms denote the *same* world. (Sommers 1993, 179–180)

The doctrine that *all true statements denote one and the same domain* (though signifying different facts) is the key to understanding why all of the 'general categorical' statements of terminized propositional logic are semantically singular" (Sommers 1993, 181),

So, how can this account of statement logic as a special branch of term logic be incorporated into the visual logic of ED? How can "terminized propositional logic" be given a graphic treatment consonant with our system for diagramming term logic in general? There are three important features of TFL (and ED) to keep in mind. The first is that the unanalyzed statements of statement or propositional logic are understood as being singular terms. Consider a sentence such as 'Aristotle is a logician', which can be diagrammed as an I categorical with the line denoting Aristotle and the line denoting logicians intersecting. But, since 'Aristotle' is a singular, uniquely denoting subject term here, it can be treated arbitrarily as having either universal or particular quantity, i.e., wild quantity. So it could equally well be diagrammed as an A categorical with the line denoting logicians properly including the line denoting Aristotle. We have seen that this means that we can simply represent a singular subject, where quantity plays no important logical role, like this using just a point. This can be graphically illustrated as follows:

Figure 3.76: How to Diagram a Singular

Secondly, we saw that the existential import of a statement might be either *implicit* or *explicit*. In the latter case, the statement's diagram contains a specified, delineated point that is not the right terminus of any term line. Thus, the intersect points of particular categoricals (as well as conjunctions) are delineated points that are not right termini; no point of universal categoricals (as well as conditionals) is a delineated point that is not a right terminus. Explicit existence for universals results from the added assumption that subject term is not vacuous; and that is represented generally by diagramming that term as both the subject and predicate term of an added particular (these differences are illustrated in Figure 3.27).

Finally, as we have seen, there is a crucial distinction to be made between particularly quantified terms (logical subjects or logical objects) when their reference is taken to be *specific* or taken to be *non-specific*. Both MPL and TFL determine the difference on the bases of the order of quantifiers: a particularly quantified term is understood as having specific reference just when it precedes but does not follow a universally quantified term. The difference is graphically illustrated by the two diagrams shown in Figure 3.62. where it must be noted that the non-terminal (i.e., left-most) point on the G line represents a specific individual (a certain girl) in the second diagram (ii) but the leftmost point on the G line in the first diagram (i) represents a non-specific individual (some girl or other). That is because this point is actually the right terminus of a line representing the complex relational term 'every boy loves ...' (−B+L), which appears analyzed (by the B line and the L arrow) in the diagram.

These three features of TFL and ED help shed light on just how propositional logic can be imbedded in (and thus seen as a special branch of) term logic. It can then be shown how ED preserves both the classical inference patterns of term logic in general but also those of propositional logic (such as *modus ponens, modus tollens,* conjunctive addition, hypothetical syllogism, etc.). In the end, it will be possible to account for the two apparent disanalogies that challenge such an incorporation. Given a statement that can be left completely unanalyzed, p, let @ denote the domain (world) "at hand" (the domain relative to which the statement is made. @ is of course singular since a statement is made relative to a *single*, specifiable, domain. While the intersection points for conjunctions are existentially explicit, conditionals have no such delineated points. Explicit "existential" import must be *made* explicit. Singular subjects always have such explicit import. Here are some simple diagrams for basic unanalyzed statement forms:

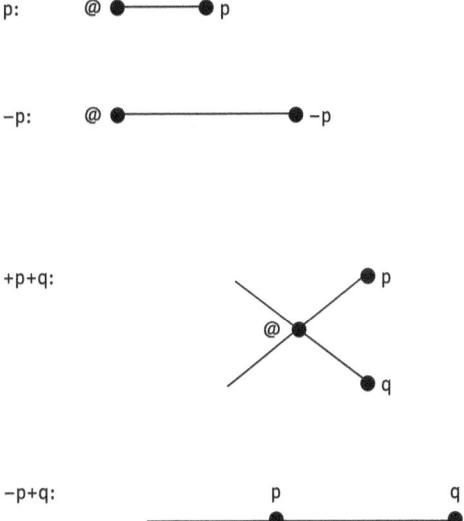

Figure 3.77: Diagrams for Sentential Logic

Note that a conditional is stated relative to a *domain of domains* (call it *D*), saying that every domain (or world) that is a p domain (a domain in which p is true) is a q domain (a domain in which q is true). We could just as well diagram it as:

Figure 3.78: The Domain of Domains for Conditionals

Note also that a conjunction (+p+q) can equally well be understood as an attribution of a conjunctive sentential term to the domain (*@+(+p+q)). Thus the two alternative diagrams for a conjunctive statement:

Figure 3.79: Alternative Diagrams for a Conjunctive Statement

Another important equivalence worth noting now is *contraposition:*

Figure 3.80: Diagrams for Contraposition

Recall that Frege had presented his system for logical notation (*Begriffsschrift*, concept-script) as a two-dimensional array of labelled lines. Moreover, he seems to have inaugurated at the same time the now-standard view that the logic of propositions is primary logic. His account of propositional logic took the functors of negation and conditionalization to be primitive. Importantly, he distinguished between the *content* of a proposition and the *judgment* (truth-claim) of it, and he distinguished them graphically with a horizontal line for the former and a small vertical line orthogonal to, and attached to, the left terminus of the horizontal line. While much of the logical theory advanced here is the result of rejecting a number of these Fregean views, it might be argued that his distinction between content and judgment can be seen in the ED representations, which indicate judgment by domain points at left termini while their absence indicates mere content. (For an insightful examination of Frege's notation for propositional logic see Schlimm, to appear.)

Since the TFL theory of logical syntax that accounts for the symbolic formulations of term logic applies as well to the logic of propositions, it is hardly surprising to see that the ED theory of graphic representation that applies to term logic applies as well to statement logic. Diagrams for some classical elementary valid argument forms (rules of inference) for propositional logic can now be constructed.

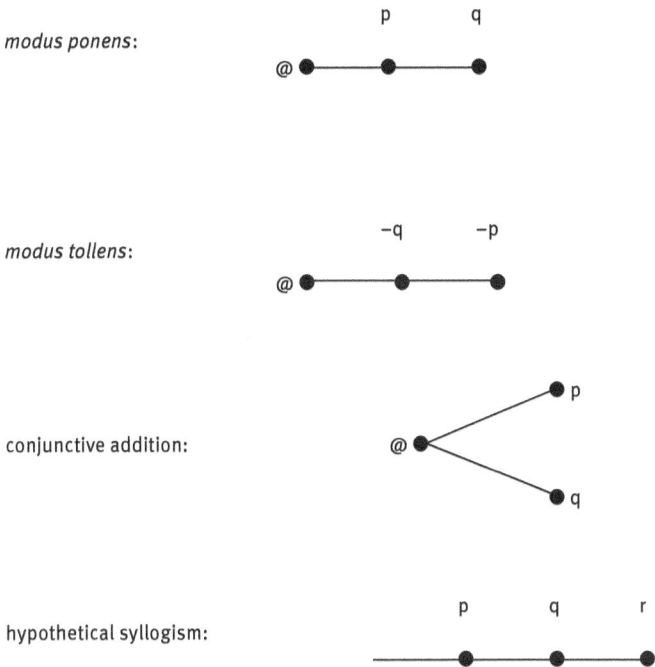

Figure 3.81: Some Elementary Rules of Propositional Logic

One can note as well that tautologies and contradictions can be diagrammed for propositional logic in just the ways they are diagrammed (Figures 3.19 and 3.20) in general (but taking the terms now to be sentential).

Just as any complex term either can be left unanalyzed or can be analyzed (both symbolically and diagrammatically, the same holds for complex sentential terms. For example, the proposition that if p then q can be left unanalyzed or it can be analyzed. The following figure represents the two results:

Figure 3.82: Unanalyzed and Analyzed Complex Terms

A few examples of deductions using this system are now in order. Consider the following argument: If p then if q then r, p, q; therefore r. We might diagram each of the three premises (step 1), then apply *modus ponens* to the first two premises (step 2), next analyze the complex term (step 3), finally, apply *modus ponens* to that result and the second premise (step 4):

Step 1:

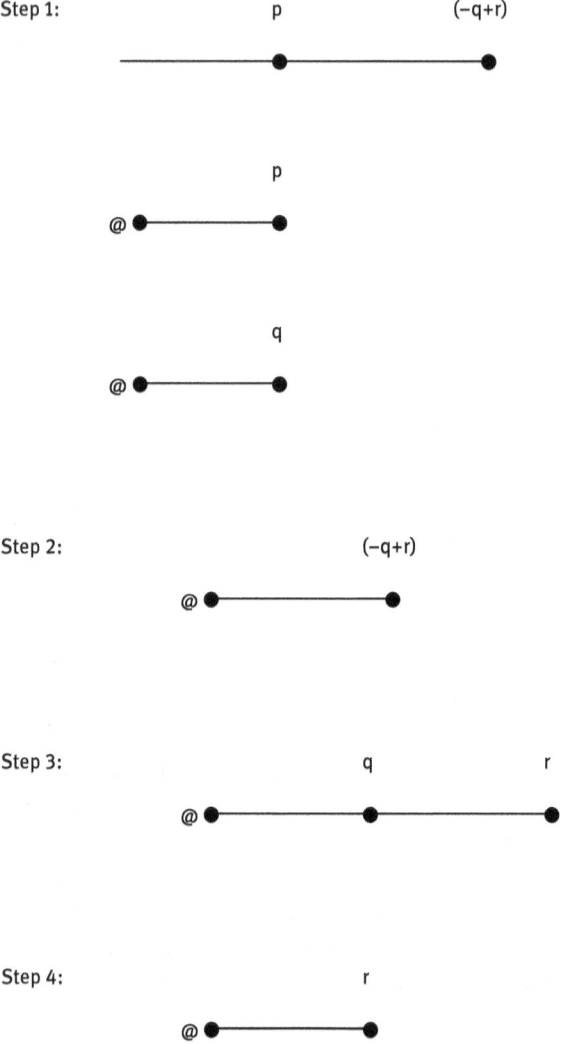

Figure 3.83: A Diagrammed Deduction

If a set of statements is consistent, its members can all be diagrammed together (i.e., no contrapiction is encountered). For example, statements of the following forms can be so diagrammed: if not r then s, if p then not r, p:

Figure 3.84: Diagram of a Consistent Set of Statements

Here is a proof that the set of four statements (if p then if not q then s, p and s, not q, not r) is not consistent by deducing a contrapiction:

136 — 3 Lines of Reason

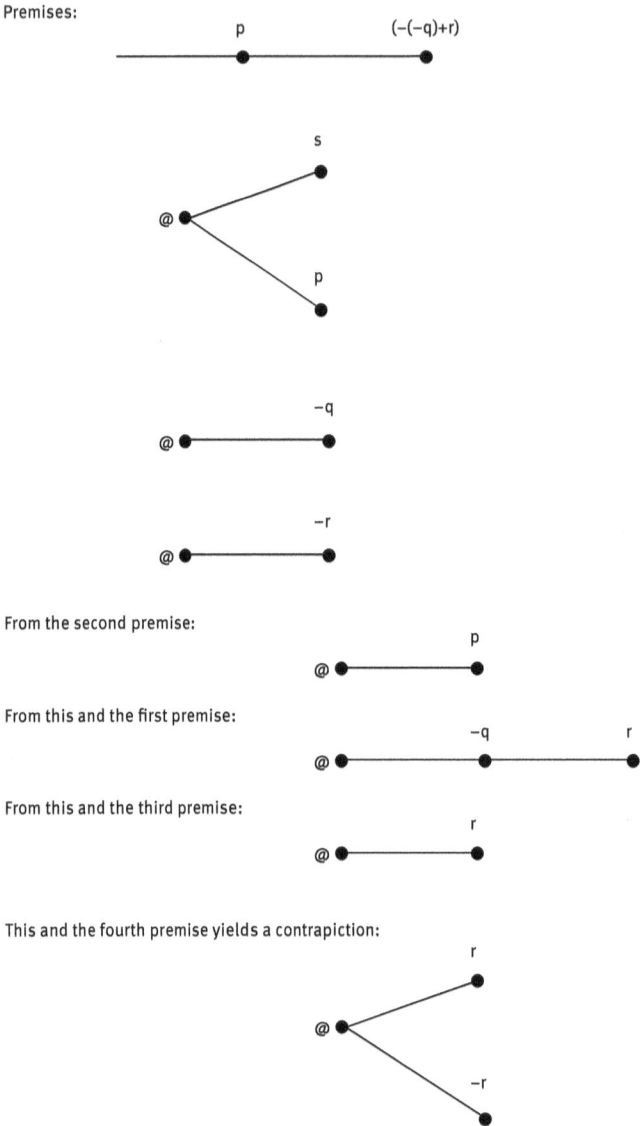

Figure 3.85: Proof of an Inconsistent Set of Statements

Note that in the first step a diagrammatic analogue of conjunctive simplification was used, viz., from an asserted conjunction any line branching from the domain point can be eliminated.

Earlier in this chapter (section 3), we have seen what the medieval logicians, Leibniz, the Port Royal logicians, the 19th century algebraic logicians, and many others have seen – the centrality of the *dictum de omni et nullo* for a logic of terms, both traditional syllogistic and TFL. It was shown there that the *dictum* amounts to a rule of substitution. In effect, the *dictum* permits the replacement of a distributed term in a sentence by the term constituting the remainder of that sentence in any other sentence in which that distributed term is now undistributed. For example, we saw that De Morgan's famous 'head of a horse' inference relies on the *dictum* in the following way. The stated premise ('Every horse is an animal') has the term 'horse' distributed. So the remaining term 'animal' can be substituted for 'horse' in another sentence which has 'horse' undistributed. The tacit, and logically innocent, premise, 'Every head of a horse is the head of a horse', has two tokens of 'horse' but only the second is undistributed. Substitution of 'animal' for 'horse' here yields the conclusion. A look at Figure 3.59 shows that H ('horse') can be ignored allowing A ('animal') to take over. The ED system of diagramming represents the application of the *dictum* whenever a middle term is ignored (see Figures 3.28–3.31). We see the *dictum* governing *modus ponens*, for example, when the consequent of the conditional is substituted for the antecedent (which must be a distributed sentential term) when that antecedent occurs in another sentence undistributed. And this is so even if the antecedent is a negative sentential term (e.g. 'if not p then q, not p; therefore, q', where the "middle" term, 'not p' is distributed in the first premise by the universal quantifier). In terms of ED, the *dictum* amounts to saying that a line segment that includes another line segment can replace that other line whenever it is not included as a proper part of another line. The fact that the *dictum*, which Leibniz called the foundational rule of mediate inference, is central to both the logic of terms and the logic of propositions, lends further weight to the conviction that the latter is a special branch of the former.

Earlier we encountered what appeared to be two disanalogies between the logic of terms and the logic of propositions. It was shown that by virtue of the fact that a sentential term denotes the unique domain of discourse relative to which it is used, the fact that this means that such terms are always singular (denoting just one thing), and the fact that when any singular term occurs quantified its quantity is wild (arbitrarily universal or particular), the disanalogies are disarmed. We will close this chapter with diagrammatic depictions of these resolutions.

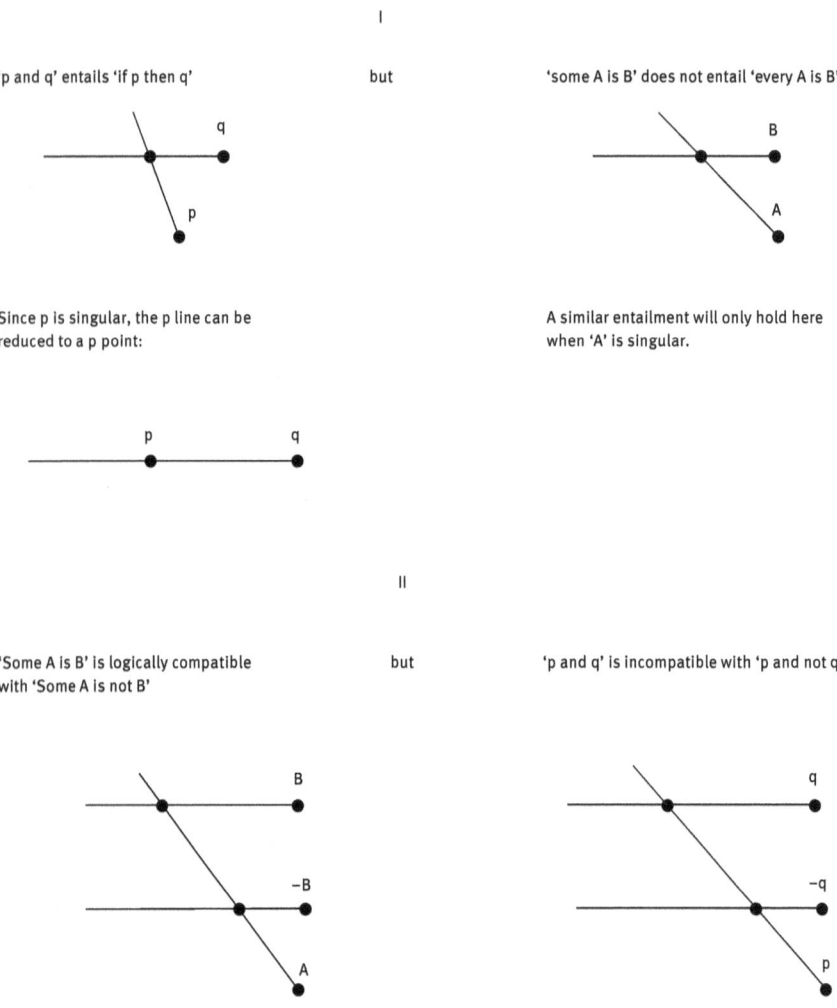

Figure 3.86: Diagrammatic Resolutions of the Two Disanalogies

In this case, since the terms are singular, the p line reduces to a p point, but there are two such points (on incompatible lines), so the diagram is a contrapiction.

4 Holding the Line

4.1 Figuring it Out

> The words or the language, as they are written or spoken, do not seem to play any role in my mechanism of thought. ... Conventional words or other signs have to be sought for laboriously only in a secondary state, when the mentioned associative play is sufficiently established and can be reproduced at will.
>
> Einstein
>
> Diagrammatic reasoning is the only really fertile reasoning.
>
> Peirce

When we say things, in some way or other, we express thoughts that we might well have kept to ourselves, left unexpressed. Among the things we say are things we say by producing statements meant to be taken as true (though they might well turn out to be false). In such cases, what are expressed are propositions, just the sort of things that are fundamental bearers of truth or falsity. As well, our rational faculties operate on propositions, primarily to make inferences, to find new information (propositions taken as true) based on other information already at hand. Our rational processes are often internal, unexpressed, but just as often they are external, expressed. Information engineering/technology aims to find ways to make the storage, manipulation, and transmission of information more efficient and faster. Logicians have contributed in many ways, both fundamentally and otherwise, to such endeavors. The results supplied by developments in modern mathematical logic have been particularly valuable here. Some logicians are also interested in simply providing an account of how we conduct our rational life in everyday settings. They build theories meant to model how we generally do it, how we naturally reason. Some of these theories rest on symbolic systems of expression, either natural or artificially created languages; other theories, as we have seen, rest on graphic systems of expression. Such graphic systems, theories of logical diagramming adequate for modelling reasoning tasks, must be able to provide models that are *efficient* but respectful of the necessary *constraints*, the limits on such models.

In the late 1980s, near the beginning of the "recent Renaissance of diagrammatology" (Danka 2016, 151; see also Stjernfelt 2007), Jill H. Larkin and Herbert A. Simon published an essay, "Why a Diagram is (Sometimes) Worth Ten Thousand Words," (Larkin and Simon 1995) that has become essential reading for diagrammatologists ever since. Larkin and Simon consider the question of what determines that one system of information representation is better as a tool for solving a given type of problem than another. Their work highlighted the importance of how one might judge the *efficacy* (see Lemon, De Rijke, and Shimojima 1999) of any system, particularly diagrammatic systems, of information representation. They consider a series of problems that are solvable by using either a system of information representation that is sentential (one-dimensional) or one that is diagrammatic (two-dimensional). In fairly assessing the comparative efficacy of *any* two information representations, the two representations must satisfy the following conditions:

> Two representations are informationally equivalent if all of the information in the one is also inferable from the other, and vice versa. ... Two representations are computationally equivalent if they are informationally equivalent and, in addition, any inference that can be drawn easily and quickly from the information given explicitly in one can also be drawn easily and quickly from the information given explicitly in the other, and vice versa. (Larkin and Simon 1995, 70)

One factor in determining how "easily and quickly" an inference can be drawn from given information is how readily one can search the given information for the elements necessary for making the inference. Larkin and Simon argue that diagrams can "exhibit efficiencies in search that would be absent from" sentential representations (Larkin and Simon 1995, 72). Other factors are how easily the elements of the given information that are necessary for the appropriate inference are recognized and, importantly, how powerful the set of rules of inference (independent of the system of representation) are that are being applied.

> Examples of this phenomenon are suggested by the everyday use of the verb "see" when no explicit visual processes are present. What is this metaphorical "seeing" and how might it connect to information-processing differences between sentential and diagrammatic representations? We speculate that this metaphor refers to inferences that are qualitatively like perceptually "seeing" in that they come about through productions with great computational efficiency. This efficiency might arise from low search and recognition costs, or from very powerful inference rules or from both. (Larkin and Simon 1995, 75)

One of the sample problems Larkin and Simon analyze is a (relatively) complex problem in plane geometry. They compare the computational costs of two solutions, one using a sentential system of representation the other using a diagrammatic system. They find these costs in the first case to be "significant problems" while the diagrammatic system's solution turns out to be "computationally very cheap, i.e., the processes of drawing and viewing the diagram" (Larkin and Simon 1995, 96). "In the sentential representation, the perceptual work of recognition is explicit and extensive; in the diagrammatic representation, it is automatic and easy" (Larkin and Simon 1995, 98). It is "automatic and easy" because "we have a mechanism – the eye and the diagram – that produces exactly the 'perceptual' results with little effort ... diagrams and the human visual system provide, at essentially zero costs, all of the inferences we have called 'perceptual'. ... this is a huge benefit. ... It is exactly because a diagram 'produces' all the elements 'for free' that it is so useful" (Larkin and Simon 1995, 99).

Larkin and Simon showed that diagrammatic systems of representation can have important advantages over sentential systems. And, just as they compared diagrammatic and sentential systems in how effective they are in the solution of various problems, similar comparisons can be made with pairs of different diagrammatic systems, in particular, systems of logic diagrams. In 1996, Atsushi Shimojima published "Operational Constraints in Diagrammatic Reasoning" (Shimojima 1996a). There he seconded Larkin and Simon's assertion that inferences based on diagrams are so useful because they give the reasoner something "for free". Shimojima calls these "free rides" (Shimojima 1996, 28). The construction of a diagram is governed by "operational constraints" that are the results of requirements determined by the rules of the diagrammatic system and which may affect how (and how well) the diagram represents information and facilitates inference. "... a free ride is essentially a case in which an operational constraint intervenes in the process of reasoning" resulting in the representation of further inferable information "*without taking any steps specifically designed for it*" (Shimojima 1996a, 32). A free ride is the "cheap" result of nothing more than an immediate, natural, visual, "perceptual inference" (Shimojima 1996a, 31). The ability of an adequate system of logic diagrams to yield free rides gives such systems an important advantage over sentential systems.

> As free rides accumulate in a longer process of derivation, the diagram site encodes a larger amount of information after relatively few steps of derivation. This explains the often-observed phenomenon that a derivation in diagrams has fewer steps than the corresponding derivation in sentences. (Shimojima 1996a, 32)

That's why, in other words, a diagram is (sometimes) worth ten thousand words.

However, with great power comes great responsibility. The power of a given diagrammatic system to provide free rides can also render the same system vulnerable to infection by instances of "overdetermined alternatives" (Shimojima 1996a, 33 ff). Unlike operational constraints that can facilitate reasoning by resulting in free rides, other constraints can impede reasoning by permitting the construction of diagrams that allow for overdetermined alternatives, diagrams that encode information that allows "us to read off unwarranted information" (Shimojima 1996a, 33). These, just like free rides, are "accidental features" of the system's operational constraints (Shimojima 1996a, 28 and 37). A simple example (Shimojima 1996a, 33–34) of overdetermined alternatives is the construction of Euler circle diagrams for the problem of determining what conclusions can be derived from the premise set: All Cs are Bs, No Bs are As, All Bs are Ds. In the diagram, the C circle is part of the B circle, which is part of the D circle. The A circle is outside of the B circle (and thus also of the C circle). But (and this is the crucial point) the A circle *must*, by virtue of the operational constraints on Euler diagrams, either be outside of the D circle or intersect with the D circle. These are the overdetermined alternatives, each of which represents information not inferable from the premise set.

While "[t]he presence or absence of free rides and overdetermined alternatives ... are two of the important factors that determine the suitedness of the particular type of diagrams to a task of reasoning ... a representation system that provides more free rides tends to impose more cases of overdetermined alternatives" (Shimojima 1996a, 47–48). So, free rides are not necessarily free lunches. In the process of formulating an adequate system of logic diagrams, what is wanted, ideally, is the optimization of free rides and the minimization of overdetermined alternatives. The overall efficacy ("suitedness") of a logic diagram system depends upon a number of factors in addition to those determined by Larkin and Simon and by Shimojima, such as its ability to represent required information (*expressiveness*) and its ability to exhibit new information (free rides) derivable by means of the inferential rules set out in the system (*inferential power*). Among other factors that have been examined is the simplification of diagrams, as well as rules of inference, by minimizing *clutter* (John, Fish, Howse, and Taylor 2006). Yet, as in the case of free rides versus overdetermined alternatives, compromise is necessary. Increasing levels of expressive and inferential powers tend to require more complex, more cluttered, diagrams. However, too much clutter tends to reduce *accessibility*.

Scientific theories generally do best when given a sufficient degree of empirical support. This is true, therefore, when it comes to theories about what constitutes adequate efficacy for reasoning with diagrams in logic. And, indeed, there have been a growing number of experimental studies in recent years designed to test such theories along a variety of parameters. Different systems of logic diagrams can be compared in terms of how they differ with respect to factors such as clutter minimization, well-matchedness/iconicity, naturalness, accessibility, and cognitive efficiency.

The term 'well-matchedness' was introduced by Gurr (Gurr 1999) to indicate the degree of homomorphism between the syntactic relations among the elements of a representation and their semantic relations. In other words, it is the measure of how well a representation, in particular a diagram, represents information, its *iconicity*. "[T]he more features there are which support a close correspondence between concept and representation the higher the iconicity of that representation" (Gottfried 2015, 6). Moktefi cites three claims central to Peirce's theory about iconicity: An icon resembles (is homomorphic with) its object, such resemblance need not be physical, and it can be more or less iconic than another icon (Moktefi 2015, 598). This third claim holds because some icons depend in part on one or another convention while other icons do not. These latter have the "highest degree of iconicity (Moktefi 2015, 607).

A representation is likely to be more effective, better suited to the task at hand, if it has what Moktefi calls *naturalness* (Moktefi 2015, 611). Consider this: the Matterhorn is a natural object. A Bierstadt painting of the Matterhorn is not natural in the same way as the mountain, but it does have a high degree of naturalness (a cubist painting would have a lower degree of naturalness). This holds for diagrammatic representations (particularly for logic) as well. In aiming to produce a highly natural diagram, it helps to make use of elements that are relatively familiar (e.g., simple plain figures such as line segments or regular closed figures rather than irregular or disconnected figures) and to exhibit relations that are more easily recognized and as immediately seen as holding among the objects represented.

Characteristics such as having minimal clutter, well-matchedness, and naturalness help to determine how readily and intuitively one can access the information represented and make appropriate inferences based on a diagram alone. They contribute to the measure of the *accessibility* of the diagram. In recent years, the assessment of, as well as the comparison of the overall accessibility and efficacy of different systems of logic diagrams have become the targets of a large and growing number of empirical studies (just some examples are: Lemon 2002; Stapleton and Masthoff 2007; Mineshima, Okada, and Takemura 2009 and 2010; Sato, Mineshima, and Takemura 2010a, 2010b, and 2011; Sato and Mineshima 2012, 2015, and 2016; Sato 2013; Cheng 2014; Sato, Wajima, and Ueda 2014; Gottfried 2015a; Blake, Stapleton, Rogers and Howse 2016; Stapleton, Blake, Burton, and Touloumis 2017; Stapleton, Jamnik, and Shimojima 2017). In most cases, these experiments involve testing significantly large groups of subjects on their abilities to perform various logic tasks, especially the relative ease and speed in interpreting different representations and performing inferences on these. A number of important results have come from such comparative investigations. Consider the following sampling:

> The primary value of an effective system of logic diagrams is its ability to relieve the reasoner of a significant amount of cognitive burden. In good measure, logical ratiocination can become little more than imaginative manipulation of what is observed (rather than just inferred), with the result being a free ride. By choosing representations with beneficial observational advantages, we reduce the need for inference and, consequently, proof systems. (Stapleton, Jamnik, and Shimojima 2017, 146).

> Categorical syllogisms can be solved better by use of Euler or Venn diagrams than by use of sentential or symbolic representations (Sato, Mineshima, and Takemura 2010a).

> Since categoricals can be taken as class inclusion/exclusion/intersection claims, such information is more easily and intuitively extracted from diagrams than from sentences. Moreover, inference can be more easily facilitated by the mere visual inspection of diagrams than sentences. ... the manipulations of diagrams [can] be spontaneously triggered without much effort, if the spatial relations holding on external diagrams are governed by... constraints that depend solely upon spatial properties of diagrams so that they are accessible even to untrained users. (Sato and Mineshima 2012, 353).

> Accessibility depends on a number of factors. One of these is a matter of how readily one can extract information of the presence or absence of an individual relative to a represented set. Euler, Venn, and Peirce's Existential Graphs offer different methods. In one recent experimental study, it was shown that representing the presence of individuals, irrespective of the associated diagram clutter, or the absence of an individual in a way that yield high diagram clutter significantly hindered task performance. By contrast, representing the presence of individuals, irrespective of the associated diagram clutter, or the absence of individuals in a low cluttered manner supports task performance. (Stapleton, Blake, Burton, and Touloumis 2017, 812).

> [L]inear diagrams function as effectively as Euler diagrams in syllogistic reasoning, and accordingly, ...the effectiveness of external diagrams in syllogistic reasoning is not due to the particular shape of set diagrams. (Sato and Mineshima 2015)

> Experimental subject groups using either linear diagrams or Euler diagrams faired equally well. Both groups performed better than groups using Venn diagrams and all groups using diagrams performed better than those using linguistic representations. (Sato and Mineshima 2012, 354)

In comparison with region based Euler or Venn diagrams, "Linear diagrams are less prone to errors and do not suffer from clutter" (Gottfried 2015a, 3). Indeed, "the construction and use of linear diagrams is simpler than the construction of region based diagrams" (Gottfried 2015a, 6). "In particular many well-formedness conditions are violated by the region-based diagrams, while linear diagrams are consistently drawn in the sense of their well-formedness conditions, though the participants are not aware of those conditions" (Gottfried 2015a, 12). "The presented experiment reveals that subjects have less difficulty with linear than with region based diagrams. ... Both the number of diagrams drawn and the number of mistakes made indicate the superiority of linear over region based diagrams" (Gottfried 2015a, 18).

4.2 Crossing the Line

> To look at a thing is very different from seeing it.
>
> Oscar Wilde

When a system of logic diagrams is maximally effective, it supports diagrams that enjoy such properties as minimal clutter, well-matchedness, and so forth. This means, for example, that such diagrams strike a careful balance between expressiveness, on the one hand, and overdetermined alternatives for inferences, on the other. But there is more. The demand for simplicity and naturalness seems to place an upper limit on the number of terms (usually four or five) that can be easily represented in a diagram. Too many terms, in effect, explode a diagram. Moreover, the expressiveness of a diagram is often limited not just by the lurking danger of permitting inferences of overdetermined alternatives. Negative terms and relational terms have tended to strictly circumscribe the expressive range, the scope of representation, of logic diagrams. Keeping all this in mind, it should be pointed out that empirical studies such as those by Sato and Mineshima (Sato and Mineshima 2012 and 2015) and by Gottfried (Gottfried 2015) strongly suggest that linear diagrams have important advantages when compared with closed figure diagrams (especially Euler and Venn diagrams).

Nonetheless, ED, the system of linear diagrams presented here, like *every* diagrammatic system is not perfect and, like any theory, must face any reasonable challenge offered. Oliver Lemon and Ian Pratt did just that (Lemon and Pratt 1998, Lemon and Pratt 1999, and Lemon 2002; see also Shimojima 1996a and 1996b; Lemon and Pratt 1997; Lemon, De Rijke, and Shimojima 1999; Stenning and Lemon 2001). The ultimate aim of the Lemon and Pratt critique is to show that representational systems that use spatial relations to model non-spatial relations (e.g., set relations) are burdened by fundamental geometric constraints that limit the expressive power of such systems. Systems of logic diagrams are particularly limited in this way. Euler and Venn circles and ED lines are targeted as typical examples of diagrams whose Achilles' heel is found in plane geometry.

> The problem with using visual languages to represent abstract (i.e. non-spatial) structures is that due to spatial constraints on the representations *not all logically possible situations can be represented visually*, so if [visual languages] are used to enumerate possible models, some logical possibilities will not be considered, leading to errors in inference if the system is used to display consequences. In short: free-rides can take graphical reasoners to the wrong destination. (Lemon 2002, 66)

Lemon and Pratt offer the following counter-example to show how the expressive power of ED is limited by geometric constraints: Consider three individuals represented by dots (points) p_1, p_2, p_3, and three sets represented line segments l_1, l_2, and l_3 (as well as their negations, not l_1, not l_2, and not l_3. Let p_1 be at the intersect of not l_1, l_2, and l_3; p_2 be at the intersect of l_1, not l_2, and l_3; and p_3 be at the intersect of l_1, l_2, and not l_3. This is diagrammed in ED as:

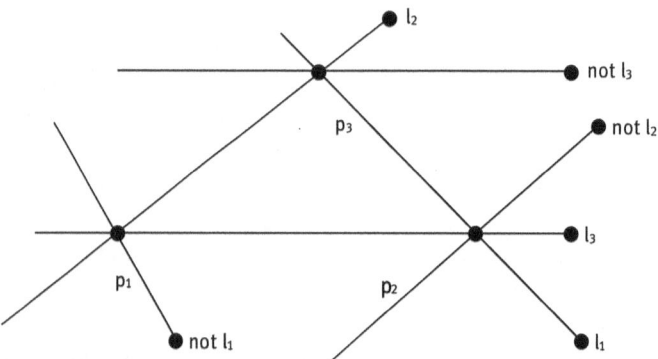

Figure 4.1: Lemon and Pratt Counter-example

Now let a fourth individual be a member of each set represented by l_1, l_2, and l_3. It turns out that this fourth individual cannot be represented in the diagram.

> To see why this augmented situation cannot be diagrammed in [ED], suppose that the new individual is represented by dot p_4. Then the four dots p_i must be distinct (since no two may lie on exactly the same [lines] l_j). It follows that the [lines] l_1 and l_2 are collinear, since they both contain the distinct dots p_3 and p_4; by similar reasoning l_1 and l_3 must be collinear, so all the [lines] lie on some common line…. (Lemon and Pratt 1998, 577)

Put simply, given four sets, A, B, C, and D, such that the intersects A∩B∩C, B∩C∩D, and C∩D∩A are all nonempty, the intersect of all four sets must also be nonempty, which need not logically follow from the three original intersect statements. The primary geometric culprit here is a result of Helly's Theorem, a theorem in combinatorial geometry that would require that for any Euclidean representation of a triple of four regions that are convex (e. g., circles or line segments) having a nonempty intersection there will be a nonempty intersection of all four regions. Thus, Euler diagrams and ED diagrams can yield overdetermined alternatives. "Of course, the expressive limitations could be bypassed here by the introduction of some new notational device" (Stenning and Lemon 2001, 46), thereby risking, however, the possibility of increased clutter.

So, the Lemon and Pratt challenge is serious, not because it applies to the Euler and ED systems alone, but because it applies generally to any diagrammatic representational system that (unlike maps, for example) goes beyond representing merely geometric properties and relations. Thus, the use of plane figures (whether closed or open) to exhibit abstract set relations can be a risky business. And yet ... diagrams (of various types) for logical reckoning continue to be used. Indeed, some, like Euler diagrams and Venn diagrams, are especially useful and are widely used in the teaching of logic as well as the elementary parts (e.g., set theory) of mathematics. Empirical studies such as the ones mentioned in the preceding section of this chapter show that non-experts faced with a variety of tasks involving logical ratiocination (particularly ones involving syllogistic inference) tend to succeed better when using diagrammatic rather than only symbolic tools. Perhaps such diagrammatic systems, though formally limited in terms of their expressive and inferential powers, enjoy quite enough power to constitute practical tools for logic. At the very least, what can be offered by the logic diagrammatologists are systems that can compare with one another (as well as with non-graphic symbolic systems) with respect to a set of reasonable criteria such as minimization of clutter, naturalness, simplicity, well-matchedness/iconicity, accessibility, and cognitive efficacy.

It would seem that a fruitful method would be one that deploys tools that allow a combination of symbolic and diagrammatic elements, a *heterogeneous logic* (see Barwise and Etchemendy 1993, 1995, 1996, 2002) idea of. This type of method has been attractive to computer science researchers. And there are other options.

ED accommodates complex relational expressions as terms that can be analyzed into their components when required, in accordance with the Principle of Relational Analysis (Figure 3.54). Compound terms could be likewise accommodated. Consider the inference: 'Some A is (both) B and C; so some A is B' (+A +(+B+C) ∴ +A+B). This can be understood as an enthymeme with the suppressed, tautological premise: 'Whatever is B and C is B' (−(+B+C)+B. We might call a generalization of this the Principle of Compound Term Analysis. So the argument would be a Darii syllogism. For simplicity's sake, label a line segment for 'both B and C' (+B+C) as 'BC'. Then our argument can be diagrammed as follows:

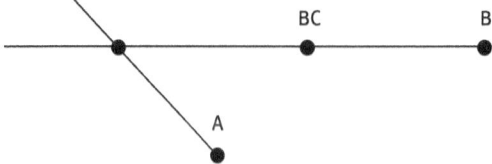

Figure 4.2: Illustration of the Principle of Compound Term Analysis

In light of this, Lemon and Pratt's counter-example (Figure 4.1) might be answered with (relabelling their numbered points and lines now as: a, b, c, d, A, B, C):

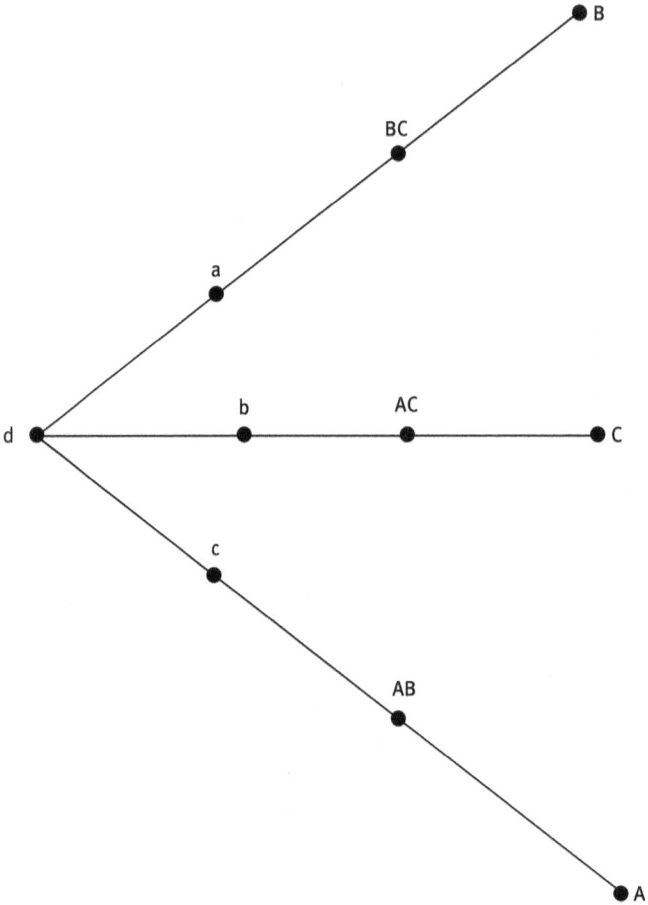

Figure 4.3: First Response to Lemon and Pratt

Another source for a response might be found in one of Peirce's diagrammatic techniques for Beta Graphs. We have seen that TFL treats so-called identity statements and categoricals with singular terms. To say that x is identical to y is merely to say that x is y and y is x. As well, following Leibniz's suggestion, singular subject terms are allowed wild quantity. ED represents any singular term as an appropriately labelled point. Figure 3.46 is helpful here. Let Aristotle be represented by a point labelled 'a'. The sentence 'a is a' could be diagrammed in at least two ways:

Figure 4.4: Line Diagrams for 'a is a'

In each case, one of the 'a' labels could be eliminated as unnecessarily redundant. That would suggest that a sentence of the form 'a is a' could be understood as well as simply 'something is a'. The result would be a graphic notation that recalls the *dots* and *lines of identity* used in Beta Graphs (see Roberts 1973, 47).

Figure 4.5: Beta Graphs for 'Something is a'

Were such graphic notations for singular terms enlisted into ED, a second answer to the Lemon and Pratt counter-example could be constructed as follows (with the same relabelling introduced in Figure 4.3):

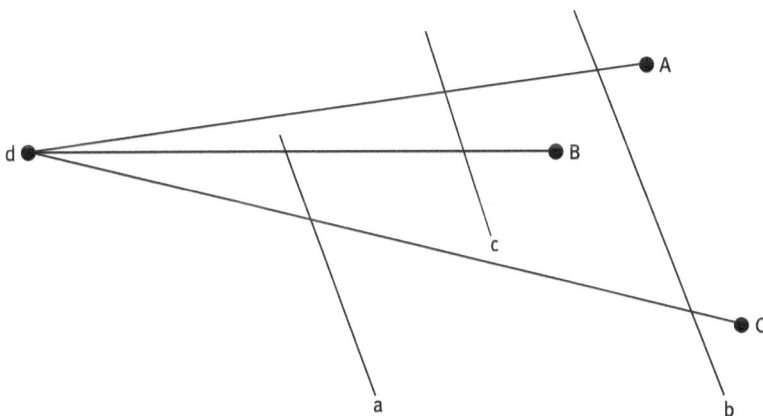

Figure 4.6: Second Response to Lemon and Pratt

Here, individual d is represented by a point (Beta dot) and the other three individuals are represented by line segments (Beta lines of identity). Such a response would require that ED be supplemented with both point and line representations of individuals as well as other accommodating emendations in both representation and inference constraints.

Keeping this in mind could lead to one more response. The reasoner might make use of a diagrammatic system that is not open-ended in terms of its range of applicability. In this regard, consider again TFL. TFL is a modern, strengthened, and extended version of traditional logic. Thus it is a symbolic logic of terms – it is syllogistic. It applies to immediate and mediate inference involving statements that can be shown to have a categorical logical form. The mediate inferences are syllogisms. Certain features are necessary for an inference to be syllogistic, especially: at least one premise must be universal and the number of particular premises must be equal to the number of particular conclusions (i.e., 0 or 1). While the application of traditional syllogistic is, in effect, confined to inferences involving only general terms, TFL incorporates singular, relational, and sentential terms.

ED is custom made for TFL.

Helly's theorem certainly holds. But the lesson to draw from the Lemon and Pratt counter-example may not be that an argument has been diagrammed that yields overdetermined alternatives because of operational constraints. Rather, the lesson could be that *no* conclusion can be drawn. The purported premises alone fail to meet one of the minimum requirements for playing the role of premises in any syllogism. All are particular (e.g., 'Something is A and B and C'). As it happens, not every consistent set of statements can be a set of premises for a valid syllogism. A set of two or more statements, each of which is particular, cannot constitute the premises of a syllogism without violating at least one of necessary requirements for being a syllogism: at least one premise must be universal and the number of particular premises must be equal to the number of particular conclusions.

ED certainly enjoys certain features sought for in a system of logic diagrams. It exhibits not insignificant levels of simplicity, minimal clutter, iconicity, naturalness, and cognitive effectiveness. Of course, such features are always relative. Given two diagrammatic systems, one might exhibit less clutter, or greater simplicity, or more iconicity, or be more natural than the other in some, but not all ways. In some cases, these differences will cancel one another, leaving our judgment about which system is more effective in the balance. When it comes to cognitive effectiveness, any difference might well tip the balance in favour of the system that enjoys the advantage in this regard. Here are a couple of examples comparing the use of Beta Graphs and ED. First, consider the two representations of 'A is greater than something greater than B'.

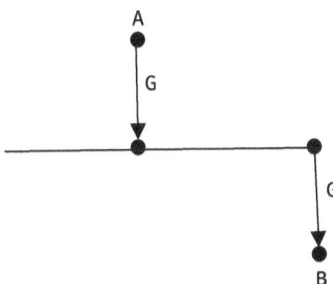

Figure 4.7: Beta Graph for 'A is greater than something greater than B'

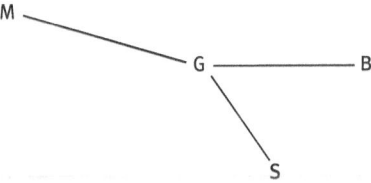

Figure 4.8: ED Diagram for 'A is greater than something greater than B'

Not much to choose from between these two. But now compare the two diagrams for 'A man gave a bribe to a senator':

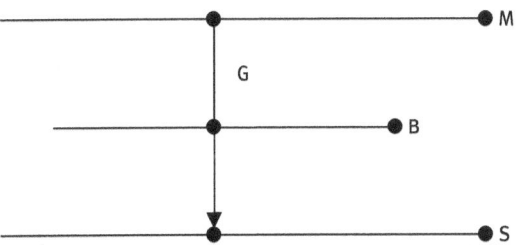

Figure 4.9: Beta Graph for 'A man gave a bribe to a senator'

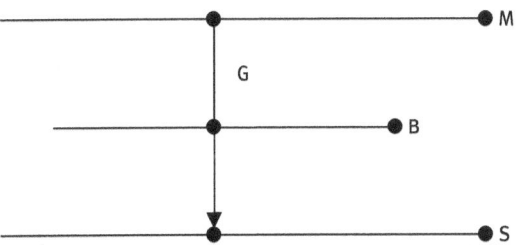

Figure 4.10: ED Diagram for 'A man gave a bribe to a senator'

The second concedes some simplicity and clutter but its exhibition of far more information gives it an important advantage over the first.

Finally, consider, without comment, the two alternative diagrams for a simple Ferio syllogism:

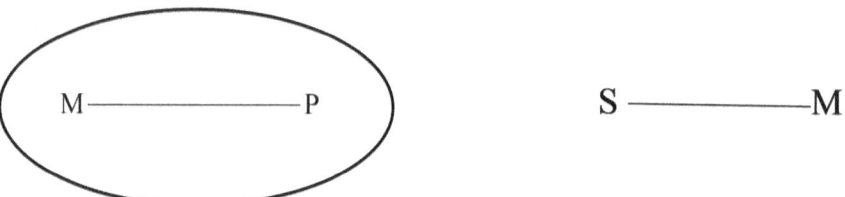

Figure 4.11: Beta Graph for Ferio

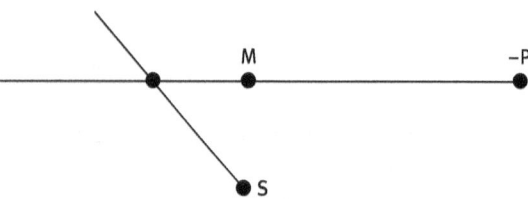

Figure 4.12: ED Diagram for Ferio

4.3 Seeing Reason

> Throughout logic, principles are borrowed from empirical psychology. The more deeply you look into the human mind in psychology, the more light you will see spread on logic.
>
> Christian Wolff

> Logic is the anatomy of thought.
>
> Locke

> The sole end of logic is to explain the principles and operations of our reasoning faculty.
>
> Hume

> I do not think I ever *reflect* in words: I employ visual diagrams, firstly, because this way of thinking is my natural language of self-communication, and secondly, because I am convinced that it is the best system for the purpose.
>
> Peirce

> There are good reasons to suppose that much, if not most, reasoning makes use of some form of visual representation.
>
> Barwise and Etchemendy

An actual reasoner must have at least some basic cognitive tools well in hand in order to carry out the tasks of effective reasoning. Long term memory, working memory, perceptual skills, language, and so forth, are usually required in some way for sound logical reckoning. When it comes to the kind of reasoning required to make valid inference, discriminate valid from invalid arguments, or solve enthymemes, some reasoners are more skilled than others. People make mistakes. Some errors are attributable to limitations on the reasoner (e.g., memory capacity, attention limits, belief biases, misperception). Other errors are attributable to the reasoning tasks themselves (e.g., premise order, subject matter, the presence of particular quantity, negation, relational terms, the so-called atmosphere effect) (see Pelletier, Elio, and Hanson 2008).

Many people, including my wife Libbey and my daughter Gaël, can put a cherry stem in their mouth, close their lips, push the stem about with their tongue, and in almost no time pull out a knotted stem. In doing this they seem to be able to form a mental image of the stem and the series of transformations required by the pushing and pulling of the tongue. Many tasks we perform by manipulating physical objects (e.g., using a wrench to remove a sparkplug from a car engine in the dark without a torch or flashlight) are like that. Even tasks that we perform not by manipulating physical objects but by assembling and modifying conceptual objects are like that. Such mental imagery is not required for all tasks but it is certainly essential for some. Is this how humans reason? Do we use internal visualizations, mental diagrams? Are diagrams the "natural language of self-communication" as Peirce found them in his own case? Are internal visualizations subject to the same constraints as external diagrams? When we *do* utilize external visual (e.g., paper-and-pencil) diagrams, are they *cognitively effective*, matching or at least augmenting our use of such mental diagrams? This last question has been a focus of several recent studies (for example, Bauer and Johnson-Laird 1993; Stenning and Oberlander 1995; Stenning, Cox, and Oberlander 1995; Stenning and Lemon 2001; Cheng, Lowe, and Scaife 2001; Blackwell 2001; Giaquinto 2007; Carey 2009; Spelke, Lee, and Izard 2010; and Hoffmann 2011). The other questions have received much attention, over many years from philosophers, logicians, diagrammatologists, information theorists, and cognitive psychologists (including those authors just mentioned).

There is no doubt that external diagrams play a major role in much of geometry. Perhaps some geometric cognition also is carried out using internal visualizations as well. And the same can be said to hold to a lesser extent for certain other areas of mathematics. Consider the number line, an old, tried and true tool for use in thinking and teaching basic arithmetic. Natural numbers don't really stand in any spatial relations with one another. Nonetheless, when engaged in numerical cognition, not thinking of them as quantities or measures of things, we seem to naturally think of them as arranged linearly with numerals representing numbers (see Nùñez 2017).

Figure 4.13: The Number Line

Conceived of as purely abstract objects, 2 is neither to the left of, to the right of, above, or below 3. Very young school children are often taught the elements of arithmetic (order, addition, subtraction) *physically*. The ultimate goal is to teach the child how to add and subtract both positive and negative numbers. In order to do this the child needs to know the difference between numerical charge (whether the number is positive or negative) and the addition and subtraction operations. Moreover, the child needs to know that numerical expressions that are not explicitly marked are always implicitly positive. Usually, large numerals are arranged linearly on large floor tiles or mats. The child is taught to begin by standing on one of them. Here, position, orientation (facing right or left), and direction of movement are all important. The student must come to understand that any task of adding or subtracting is initiated by the first expression that occurs in the operation. For example, in adding 3 and 2, the child must see both that the 3 represents a positive number and that it locates her on the number line (so she stands at +3). Next, the charge (whether explicit or not) on the second term determines her orientation: facing right for positive or left for negative charge. Finally, the operation sign indicates the direction of movement. Addition means moving forward while subtraction means moving backward. Subtraction does not mean that the child must move to the left. In fact, subtraction can involve movement either to the left or to the right. What matters is the orientation of the child and her backward motion, which is to the left only when she is facing right. With practice, the child soon learns that the addition and subtraction of numbers (positive or negative) is, first, fun and then easy – lessons that, unfortunately, are often lost in the fullness of time (for more on this teaching method see Boulet, Francavilla, and Englebretsen 1994).

Students are usually taught these things without recourse to such physical training. They are simply introduced (or re-introduced) to them using the *external visual* number line indicated in Figure 4.13. Eventually, if things go well, the external line can be replaced by an *internal visual* number line. Presumably, Peirce would have felt comfortable at this stage. In contrast, many (most?) mathematicians and logicians would count all of these number lines (physical or externally visual, or even internally visual) as merely heuristic or learning aids, inessential when carrying out the true business of mathematics, where only pure cognition, *internal* and *non-visual*, is required.

Perhaps. The question remains, however: Can we think in terms of something like mental diagrams? Are they cognitively effective? Indeed, more generally, how do we actually reason? Sommers' work on the development of a viable, powerful logic of terms, TFL, was in part motivated by his conviction that a system of formal logic such as this would serve as a better descriptive model of how we naturally reason. This was so, in part, because the syntax of TFL is meant to hew much more closely than that of MPL to the syntax of the natural language we ordinarily use. Importantly, when ordinary people in ordinary situations reason they do so in their own natural language and not in any regimented artificial language, not the language of MPL, not even that of TFL. So, how *do* we naturally reason? W.S. Cooper has given a fairly radical answer.

"How do humans manage to reason?" (Cooper 2001, 4). Cooper's answer is that rationality is simply the result of human evolution. "The laws of logic are not independent of biology but implicit in the very evolutionary processes that enforce them. The processes determine the laws" (Cooper 2001, 2). It turns out for Cooper "that biology is all there is to logic" (Cooper 2001, 173). The "general evolutionary tendency to optimize fitness turns out to imply, in and of itself, a tendency for organisms to be rational" (Cooper 3001, 9). "A fit cogitator will avoid ... irrationalities as a primate avoids snakes and for the same reason: A tolerance of irrationalities, like a tolerance for snakes, is unstable and invadable by less tolerant strategies" (Cooper 2001, 80–81). "A law of logic becomes a factual generalization in a scientific theory of a sort not fundamentally different from any other" (Cooper 2001, 192). What he terms "classical logic," which assumes that basic principles of reasoning are necessary and *a priori*, "is unfit for human use" (Cooper 2001, 180). It seems to follow that logic is relative (since biology is relative). "The laws of logic are never absolute but always relative to the contingencies of some underlying evolutionary model. ... Different earthly species can have different logics" (Cooper 2001, 179). It would seem that Cooper's knowledge of evolutionary biology surpasses his understanding of the nature of formal logic.

Nonetheless, it might well be the case that our ability to reason is, after all, innate. Noam Chomsky, the father of the thesis that our knowledge of the deep grammar of the universal language that underlies our natural languages is innate, holds as well that our logical ability (in particular as modelled by MPL) is innate (Chomsky 1980, 165, and Chomsky 1981, 34; see Sommers 2008b, 120–121, for a response). Chomsky's innateness claim about language is the foundation for an explanation of how very young children acquire relatively sophisticated levels of linguistic capacity so early and so quickly. The idea that rational capacity is innate is often referred to as *nativism*. A stark expressions of it is: "It is nonsense to think that basic logical skills can be learned" (Macnamara 1986, 28). Other prominent expressions of nativism are: (Wetherick 1989, Rips 1994 and 1995, Crain and Khlentzos 2008 and 2010). While rational ability might be innate, it seem to be exhibited and developed more slowly than linguistic ability. If there *is* an innate, natural, intuitively applied logic, then it must offer a logical syntax that is intimately related to the syntax of the natural language sentences one uses in reasoning deductively, and it must be relatively simple, certainly simpler than the kinds of logic offered in today's logic textbooks (Sommers 2008b, 116). Even if our rational capacity is not innate, the fact that we come to use it relatively effectively at least by puberty suggests that it is the result of a long and remarkable evolutionary process. "In the course of becoming rational animals, human beings have had eons to develop language and to hit upon a method of deductive reasoning with the sentences of their natural language" (Sommers 2008a, 5). Like Cooper (though far more moderately), Sommers sought an empirically informed account of how we actually reason, concluding that it must be found in a "cognitively veridical" logic, one descriptive of our everyday reasoning (Sommers 2008a, 6). "To fulfill its traditional mission of exposing the 'Laws of Thought' Logic must directly address the cognitive puzzle of how untutored human beings reason so well" (Sommers 2008a, 7).

> Any satisfactory report of what is going on in a piece of ratiocination must take into consideration the fact that we intuitively reason with the natural ... sentences of our native language, that we do so with extraordinary celerity, efficiency and confidence, and that mostly, our reasoning is valid. In looking for a psychologically veridical and logically sound account of how we do these things, we learn a lot about logic itself. (Sommers 2008b, 122)

Sommers' solution to the cognitive puzzle is his "cognitive hypothesis" that in our everyday deductive reasoning we exploit the plus/minus characteristics of terms in accordance with the principles of TFL (Sommers 2008b, 124–125, and Sommers 2008a, 6). "Term Functor Logic is thus revealed as the natural logic of our everyday ratiocinations. It is how our deductive intuitions work. Of course, intuition is not infallible; we often reckon wrongly" (Sommers 2008b 132–133). Since this cognitive hypothesis is an empirical conjecture, it must (eventually) rest on a body of adequate evidence. "It is reasonable to expect that when neuroscience emerges from its infancy we will get discriminate information about what happens in the brain as we [reason deductively]" (Sommers 2008b, 116; see also Sommers 2008a, 128).

If one accepts the cognitive hypothesis concerning deductive logic, need it apply only to reasoning carried out symbolically? We have returned now, in effect, to the questions raised above: Can we reason deductively using either internal or external diagrams? Can the use and manipulation of such diagrams for deductive reasoning be cognitively effective, cognitively veridical? More to the point: *Do* we ever reason with icons rather than symbols?

> Strictly speaking, the eyes are part of the brain. From the cognitive point of view, however, the perceptual apparatus is fairly autonomous. The sorts of things it does are fairly well insulated from the forms of reasoning we engage in at a conscious level. Nevertheless, the perceptual system is an enormously powerful system and carries out a great deal of what one would want to call inference, and which has indeed been called perceptual inference. (Barwise and Etchemendy 1996, 25)

One might quibble of course with just how well perceptual inference is actually "insulated from the forms of reasoning we engage in at a conscious level."

4.4 So, How Do We Reason, After All?

> [W]e must conceive that which is generated through sense-perception in the sentient soul, and in the part of the body which is its seat – viz. the affection the state whereof we call memory – to be some such thing as a picture.
>
> Aristotle

> Wir Machen uns Bilder der Tatsachen. ...
> Das logische Bild der Tatsachen ist der Gedanke. ...
> "Ein Sachverhalt ist denkbar", heißt: Wir können uns ein Bild von ihm machen.
>
> Wittgenstein

> [T]here are no such objects as mental pictures.
>
> Ryle

As we all know, Frege rejected any notion that logic aims to *describe* how we think, our thoughts. Instead, logic is solely concerned with the principles governing Thoughts (*Denken*). It must *prescribe* how one ought to think. The laws of logic "prescribe universally the way one ought to think if one is to think at all" (Frege 1967, 12). Our *thoughts* are subjective, private, mental affairs, targets of psychology proper. In contrast, *Thoughts* are not mental; they are objective, public, and shareable, the proper targets of logic. Logicians who confuse the two are guilty of "psychologism" and taint logic with psychological concerns. When Sommers advocated for a cognitively veridical, realistic logic as a fertile area of investigation for cognitive psychology, he noted that there has been an impediment to this due to the kind of stand Frege took against psychologism in logic. Nonetheless, as Sommers wrote, "There are signs that anti-psychologist bias of modern logic is on the wane (Sommers 1983d, 44). Of course, 'psychologism' can mean different things to different logicians (see Pelletier, Elio, and Hanson 2008). Much of the force of Frege's anti-psychologism is due to the failure of other logicians, who focused on an individual's thoughts (rather than Thoughts), to account for the objectivity and necessity of laws of logic. Recently, Vanessa Lehan-Streisel has defended what she calls "social psychologism" as opposed to the individual or personal psychologism that Frege had targeted (Lehan-Streisel 2012). She argues that the most objectivity and necessity that one can expect to find is located among the various ways that groups of people communicate their reasonings with one another no matter the natural language being used. She writes:

> I take the justification of good inference to be necessarily based in the meaning of logical terms as they are used in natural language. ... The problem with the anti-psychology position is that one of our concerns, as philosophers, is to explain good and bad inference. It is this "actual reasoning" that is of concern to us (among other things, of course). (Lehan-Streisel 2012, 579)

> It is not enough to have a normative theory outlined, we need also to have some idea of how people come to know or to use logical terms and inferences. ...there is good evidence already to believe that the central rules of good reasoning will be shared across languages. (Lehan-Streisel 2012, 580–581)

The kind of *reasoning* these philosophers, logicians, mathematicians, and cognitive psychologists have all had in mind of course is *deductive reasoning*. At the very beginning of his two volume *Analytics*, Aristotle tells us that his inquiry is concerned with *demonstration*. It turns out that a demonstration is a proof, a *deduction* from known premises to an (un)known conclusion. A deduction (viz., a syllogism) is an argument whose validity (having the conclusion necessarily follow from the premises) is either immediately obvious or is an argumentation consisting of a chain of such arguments. Deduction, syllogistic, therefore, must be dealt with first. "The reason why we must deal with the syllogism before we deal with demonstration is that the syllogism is more universal; for demonstration is a kind of syllogism, but not every syllogism is a demonstration" (*Prior Analytics* 25b26–31). Since *Prior Analytics* presents the deductive logic of syllogisms, demonstration is reserved for treatment in *Posterior Analytics* (see Corcoran 2009).

We have seen that some psychologists of reason advocate the nativist view that the capacity to reason is innate rather than learned. Perhaps the most important researcher in the field of the psychology of reason is Philip Johnson-Laird (see Johnson-Laird 1983, Johnson-Laird and Byrn 1991, Bauer and Johnson-Laird 1993, Johnson-Laird 2002 and 2010). Among Johnson-Laird's most important claims is that the capacity to reason is dependent on the ability to form and manipulate *mental models*. A mental model can be derived from remembered perceptions, imagination, or even entertained linguistic expressions. In effect, one constructs a mental picture representing some possible situation in the real or imagined world and then reasons by manipulating such a model or combinations of such models.

> Logically untrained individuals tend to reason by constructing models of the situations described or depicted in the premises, and the effect of diagrams on the process is the first practical application of the theory of mental models. (Bauer and Johnson-Laird 1993, 378)

Advocates of such an account of reasoning have been termed "visualisers" in contrast with "verbalisers" (Stenning and Lemon 2001). The verbalisers don't deny that we can use mental models, but hold that such an ability is not prior to the ability to reason with symbolic, verbalisable mental representations. As one prominent verbaliser has said, "The basic unit is a mental sentence" (Rips 1995, 316; see also Rips 1994, Braine 1978, and Braine and O'Brien 1998). As it happens, Stenning and Lemon, in attempting to illustrate how logic and psychology "are inseparable in an account of thinking with diagrams" (Stenning and Lemon 2001, 58), showed empirically that student subjects "are adept at strategically choosing when to translate between modalities (from sentential to graphical or from graphical to sentential). In fact, the 'verbalisers' are characterised by a tendency to translate immediately into the graphical modality, and to fail to translate in the opposite direction appropriately" (Stenning and Lemon 2001, 41–42).

Constructing and manipulating something like a mental model, an internal logic diagram, can be used in deduction. "Clearly, such a knowledge representation can make some inferences, and yet the inferences inhere in the knowledge representation, including the construction and retrieval processes, and do not require a separate proof procedure" (Lindsay 1995, 115). "No claim is made that [such inference] methods can in principle do things that logic cannot. It is likely, however, that they can do some things *more efficiently*, and that is the crux of the matter" (Lindsay 1995, 121). This view is reinforced by Larkin and Simon's claim "that mental images play a role in problem solving quite analogous to the role played by external diagrams. (Larkin and Simon 1995, 105), adding that the creation of mental images "employs inference processes like those that make information explicit in the course of drawing a diagram" (Simon and Larkin 1995, 106). The fact is that we do often, and quite naturally, carry out our mental acts of ratiocination by constructing and manipulating internal diagrams. By now of course, it should be obvious that we also often do so as well by using external diagrams. Indeed, it is our use of the latter that offers a "perspective on the nature of internal representations used when reasoning without explicit external graphics" (Stenning and Oberlander 1995, 126). Stenning and Oberlander go on to cite several empirical studies that support the claim that "graphical representations are easier to process *for any reasoning system*" (Stenning and Oberlander 1995, 126). One example helping to confirm this claim is offered by Hamami and Mumma who cite a number of other empirical studies revealing that people untutored in geometry have a wide range of cognitive intuitions, basic concepts and capacities relevant for geometrical reasoning (Hamami and Mumma 2012, 18 – 19). In approaching an understanding of how we use internal diagrams in the process of logical deduction, it should always be kept in mind that reasoning in general is an activity dependent on a capacity, a *knowing how*. It can be done well or poorly. One need not explicitly *know* the rules of deduction (i.e., be able to articulate, or even cite, the rules) but, given that a person does know how to deduce, her deductions are logically correct. "These rules are of course explicit representations of implicit reasoning practices. ... This is not to say that ordinary reasoners would necessarily recognize them as such" (Legg 2013, 14). Reasoning, then is like juggling. If you actually juggle it can be said that you know how to juggle, whether you know in any sense the principles of juggling or not.

Since reasoning is like juggling it takes instruction and practice. Presumably, the reasoner was previously fitted for similar instruction and practice, was capable of learning to reason, had the capacity to do so. If reasoning is a natural, innate capacity for humans, then cognitive science is not the only place to look for insights and evidence about what it is and how we acquired it. Biology is also an obvious and promising place to look. Perhaps surprisingly, even ancient biology can be of help. The Cartesian idea that living organisms are machines, mechanical devices governed by the laws of physics, is still alive and well in many quarters. While most biologists are reluctant to go so far as to hold, with Descartes, that humans differ from all other animals by being somehow (perhaps divinely) fitted with an immaterial *mind*, they nonetheless often agree with many non-biologists that humans are special kinds of machines, namely computing machines with functional programs playing the mental role. The Cartesian *dualism* account of humans supplanted a much older view. If Descartes was a physicist-epistemologist, then Aristotle was a biologist-ontologist. He took different kinds of material objects (e.g., artifacts, plants, animals) as having different natures (or *forms*). In the case of a living organisms this is its *psuché*, its *psyche*. A psyche is an organism's capacity to do certain things (e.g., grow, perceive, think, etc.). While Descartes said he was a thing that thinks (a mind), Aristotle said that he was a rational animal. For Aristotle, psychology, the study of the psyches of living organisms, was a part of biology. A key element of such a study was the conception of these organisms as *units*. A plant, animal, or human is not a combination of matter with psyche. The latter is a form. Neither matter nor forms are *substances* – unified things, formed matter/enmattered form. Only substances can change, act, exercise their capacities. "It is doubtless better to avoid saying that the soul pities or learns or thinks, and rather to say that it is the man who does this with his soul" (*De Anima* 408b13–15). My eyes don't see; I see with my eyes. My hands don't catch the ball; I catch it with my hand. This is Aristotelian *monism*.

But how and why have humans, unlike other animals, evolved to have the capacity to reason? How is it that we are rational animals? Recently, in a study revealing a great deal of sound scientific research as well as rich philosophical insight, Harry Smit (Smit 2018) convincingly shows how much light can be shed on such questions by the integration of neo-Aristotelian monism with post-Darwinian biology. The result is an account of the evolution of human rationality as the natural development of a suite of capacities, psychic powers (in the serious sense of that phrase) attributable not to our minds, or brains, or brain parts, but to us, as living human organisms. "The mental, following the neo-Aristotelian conception, is *manifest* or expressed in behaviour and is therefore directly observable" (Smit 2018, 294). Just as our capacity to move and perceive is revealed by what we do, so is our capacity to reason, which is revealed (mostly) by what we do with words, what we say. Such doings are intentional. We do them for a reason, we aim at some result. I raise my hand to get the teacher's attention.

> Language evolution freed humans from the constraints imposed by genetic evolution for it enabled us to reason and give reasons. ... But this developmental pattern (from doing things with words to reasoning and giving reasons) evolved during our (evolutionary) history. In neo-Aristotelian terms: the rational psyche evolved out of the sensitive psyche when humans became *sensitive* to the reasons for acting and responding. (Smit 2018, 295)

This transition from the sensitive to the rational was the result of "genetic variations in the expansion of our vocalization powers and in socially guided learning" (Smit 2018, 310).

Very young children babble, then replicate what they take to be babble from others, then use words. The use of words can be correct or incorrect (thus requiring some teaching) – it becomes normative. At the age of 3 or 4 children combine words, even in some sophisticated or complex ways (tensed, conditional, using pronouns, etc.). They then come to think counterfactually, modally, etc. They can conceive of nonexistents. They imagine. They reason. "[C]hildren acquire the skill to refer to entities outside the communication-situation, to talk about persons who are not present, to focus their attention on something unrelated to current needs and wishes, and so on and so forth. They have acquired the rational psyche" (Smit 2018, 310). Our rational psyche, our capacity to reason and to give reasons evolved. It is not something done by a Cartesian mind. It is done by us.

Finally, cognitive neuroscientists and evolutionary psychologists might want to attribute to the brain or mind something like *information processing*. Such an idea finds its genesis in Descartes of course. But the idea fails to explain the evolution of language and reasoning. We did not somehow evolve a new kind of brain, brain part or mind (Smit 2018, 311). Our rational psyche is the result of natural evolution. Rationality is natural.

5 Linear Diagrams and Non-Classical Quantifiers

5.1 Introduction

In this chapter we attempt to achieve two goals: *i)* to meet the advances of Sommers (1967, 1982, 2000) and Englebretsen (1987b, 1996, 2000) (namely, a plus-minus algebra for syllogistic) together with the developments of Peterson (1979) and Thompson (1982) (namely, an extension of syllogistic with "most", "many", and "few"); and *ii)* offer a diagrammatic device able to capture such meeting. These goals should be of relative interest for two reasons: *i)* the plus-minus calculus provides an algebraic approach for syllogistic that, alas, does not cover cases of common sense reasoning involving non-classical quantifiers such as "most", "many", or "few", whereas the syllogistic extended with these extra quantifiers comprises a wide range of common sense inference patterns but lacks an algebraic procedure; and *ii)* there is no diagrammatic system able to capture these ideas, as far as we are aware (Cf. Gardner 1958, Moktefi & Shin 2012).

So, given this state of affairs, in this study we offer, respectively, *i)* an extension of syllogistic that includes a broad range of inferential patterns but with the virtues of an algebraic approach and *ii)* a diagrammatic device designed to represent such an extension. To reach these results we briefly present two frameworks around syllogistic (§2), then we introduce our contributions (§3–4) and, at the end, we succinctly mention some possible uses of these extensions (§5).

5.2 Two Frameworks for Syllogistic

5.2.1 General aspects of syllogistic

Syllogistic (SYLL) is a term logic that has its origins in Aristotle's *Prior Analytics* (1989) and deals with the consequence relation between categorical propositions. A *categorical proposition* is a proposition composed by two terms, a quantity, and a quality. The subject and the predicate of a proposition are called *terms*: the term-schema S denotes the subject term of the proposition and the term-schema P denotes the predicate. The *quantity* may be either universal (*All*) or particular (*Some*) and the *quality* may be either affirmative (*is*) or negative (*is not*). These categorical propositions are denoted by a *label*, either a (universal affirmative, SaP), e (universal negative, SeP), i (particular affirmative, SiP), or o (particular negative, SoP) that allows us to determine a sequence of three propositions called *mood*. A categorical syllogism, then, is a mood ordered in such a way that two propositions are premises and the last one is a conclusion. Within the premises there is a term that appears in both premises but not in the conclusion. This special term, usually denoted with the term-schema M, works as a link between the remaining terms and is known as the middle term. According to the position of this last term, four *figures* can be set up in order to encode the valid syllogistic moods or syllogistic patterns (Table 5.1).[1]

Table 5.1: Valid syllogisms

Figure 1	Figure 2	Figure 3	Figure 4
aaa	eae	iai	aee
eae	aee	aii	iai
aii	eio	oao	eio
eio	aoo	eio	

5.2.2 The TFL framework: the plus-minus algebra

Sommers (1967, 1982, 2000) and Englebretsen (1987b, 1996, 2000) developed a plus-minus algebra, *Term Functor Logic* (TFL), that deals with syllogistic by using terms rather than first order language elements such as individual variables or quantifiers.[2] According to this algebra, the four categorical propositions can be represented by the following syntax:[3]

- SaP := -S+P = -S-(-P) = -(-P)-S = -(-P)-(+S)
- SeP := -S-P = -S-(+P) = -P-S = -P-(+S)
- SiP := +S+P = +S-(-P) = +P+S = +P-(-S)
- SoP := +S-P = +S-(+P) = +(-P)+S = +(-P)-(-S)

---

[1] For sake of brevity, but without loss of generality, here we omit the syllogisms that require existential import.
[2] Since we can reason without first-order language elements, such as individual variables or quantifiers, is not news (Cf. Quine 1971, Noah 1980, Kuhn 1983), but Sommers' logical project has a wider impact: that we can use a logic of terms instead of a first order system has nothing to do with the mere syntactical fact, as it were, that we can reason without quantifiers or variables, but with the general view that natural language is a source of natural logic (Cf. Sommers 2005c, Moss 2015).
[3] We mainly focus on the presentation by Englebretsen (1996).

Given this algebraic representation, the plus-minus algebra offers a sound, complete, and simple method of decision for syllogistic: a conclusion follows validly from a set of premises if and only if *i)* the sum of the premises is algebraically equal to the conclusion and *ii)* the number of conclusions with particular quantity (viz., zero or one) is the same as the number of premises with particular quantity (Englebretsen 1996, p. 167). Thus, for instance, if we consider a valid syllogism from Figure 5.1, we can see how the application of this method produces the right conclusion (Table 5.2).

Table 5.2: An aaa-1 type syllogism

	Proposition	Representation
1.	All dogs are animals.	-D+A
2.	All German Shepherds are dogs.	-G+D
⊢	All German Shepherds are animals.	-G+A

In the previous example we can clearly see how the method works: *i)* if we add up the premises we obtain the algebraic expression (-D+A)+(-G+D)=-D+A-G+D=-G+A, so that the sum of the premises is algebraically equal to the conclusion and the conclusion is -G+A, rather than +A-G, because *ii)* the number of conclusions with particular quantity (zero in this case) is the same as the number of premises with particular quantity (zero).

This algebraic approach is also capable of representing relational, singular, and compound propositions with ease and clarity while preserving its main idea, namely, that inference is a logical procedure between terms. For example, the following cases illustrate how to represent and perform inferences with relational (Table 5.3), singular[4] (Table 5.4), or compound propositions[5] (Table 5.5). For a brief but systematic explanation of the rules employed in what follows *vide* Appendix A.

---

[4] Provided singular terms, such as *Socrates*, are represented by lowercase letters.
[5] Given that compound propositions can be represented as follows, $P:=[p]$, $Q:=[q]$, $\neg P:=-[p]$, $P \rightarrow Q:=-[p]+[q]$, $P \wedge Q:=+[p]+[q]$, and $P \vee Q:=-[p]-[q]$, the method of decision behaves like resolution (Cf. Noah 2005). Also, for a better understanding of the difference between a reasoning with singular terms, like the example in Table 5.4, and a propositional argument involving a conditional statement, like the example in Table 5.5, consider the analysis provided by (Pereira-Fariña 2014).

Table 5.3: A reasoning with relational propositions

	Proposition	Representation	Rule
1.	Some horses are faster than some dogs.	+H+(+F+D)	P
2.	Dogs are faster than some men.	-D+(+F+M)	P
3.	That which is faster than what is faster than some men, is faster than some men.[6]	-(+F+(+F+M))+(+F+M)	P
4.		+H+(+F+(+F+M))	DON 1,2
⊢	Some horses are faster than some men.	+H+(+F+M)	DON 3,4

Table 5.4: A reasoning with singular propositions

	Proposition	Representation	Rule
1.	All men are mortal.	-M+L	P
2.	Socrates is a man.	-s+M	P
⊢	Socrates is mortal.	-s+L	DON 1,2

Table 5.5: A reasoning with compound propositions

	Proposition	Representation	Rule
1.	If P then Q	-[p]+[q]	P
2.	P	+[p]	P
⊢	Q	+[q]	DON 1,2

5.2.3 The SYLL⁺ framework: the extra quantifiers

Peterson (1979) and Thompson (1982) developed extensions for syllogistic (SYLL⁺) by adding some extra quantifiers, namely, "most" (for majority propositions), "many" (for common propositions), and "few" (for predominant propositions).[7] So, this framework adds the next propositions: p is the predominant affirmative (*Few* S *are not* P), b is the predominant negative (*Few* S *are* P), t is the majority affirmative (*Most* S *are* P), d is the majority negative (*Most* S *are not* P), k

---

[6] Or in other words: the relation *faster than* is transitive.
[7] We mainly focus on the presentation by Thompson (1982).

is the common affirmative (*Many* S *are* P), and g is the common negative (*Many* S *are not* P).

According to Thompson (1982, p. 76), these new quantifiers may be understood in any of three senses: minimal, maximal, and exact. A quantifier in the minimal sense is understood to be saying "at least" or "no less than" the quantity named. A quantifier in the maximal sense is understood to be saying "only" or "no more than" the quantity stated. A quantifier in the exact sense combines the minimal and maximal senses, so that it says "no more nor less than". Given this taxonomy, Thompson uses "many" and "most" in a minimal sense, just as "some" is traditionally understood as meaning "at least some (and possibly all)". Meanwhile, Thompson restricts "few" to its maximal sense as to mean "no more than few (if any)": this results in the notion that "few" must be understood to be making a denial, and this explains why the proposition "*Few* S *are* P" is regarded as the predominant negative, whereas "*Few* S *are not* P" is taken as the predominant affirmative. All these semantic remarks result in an extended square of opposition (Figure 5.1).

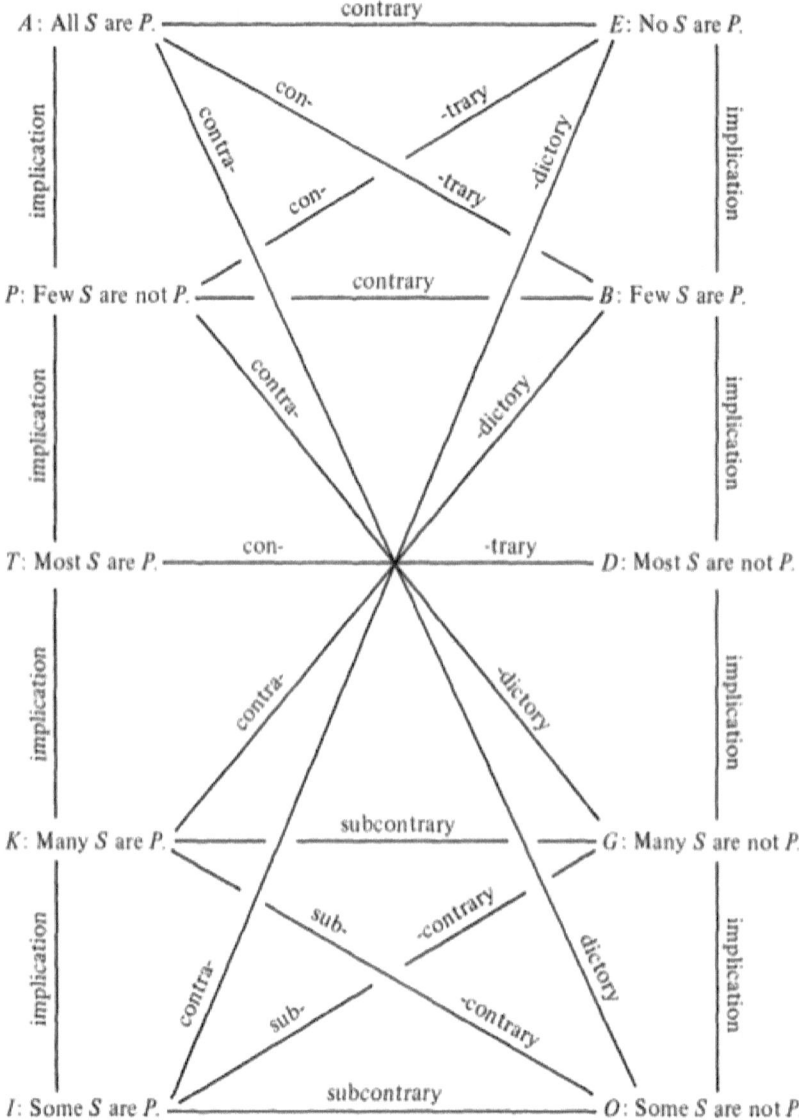

Figure 5.1: Extended Square of Opposition Adapted From (Thompson 1986, p. 77)

Given these new propositions, SYLL⁺ requires some basic assumptions of distribution: any universal proposition distributes its subject; any negative proposition distributes its predicate; any predominant, majority, or common proposition distributes its subject if and only if its subject is the minor term. With these basic assumptions, the SYLL⁺ framework provides the next rules of validity:

1) Rules of distribution.
   a) The middle term must be distributed in at least one premise.
   b) Any term which is distributed in the conclusion must also be distributed in the premises.
2) Rules of quality.
   a) There must be at least one affirmative premise.
   b) If the conclusion is negative, there must be at least one negative premise.
   c) If there is a negative premise, then the conclusion must be negative.
3) Rules of quantity.
   a) If there is a predominant premise, the conclusion may not be universal.
   b) If there is a majority premise, the conclusion may not be universal or predominant.
   c) If there is a common premise, the conclusion may not be universal, predominant, or majority.

It is clear this framework allows us to extend syllogistic as to cope with a wide range of common sense reasoning patterns, both valid (Table 5.6) and invalid (Table 5.7: we denote the fact that the conclusion does not follow by "⊬"). As expected, the addition of p, t, k, b, d, and g increases the number of valid syllogistic moods (Table 5.8).

Table 5.6: A valid reasoning: aat-1

	Proposition	Representation
1.	All humans are mortal.	HaM
2.	All Greeks are humans.	GaH
⊢	Most Greeks are mortal.	GtM

Table 5.7: An invalid reasoning: tta-1

	Proposition	Representation
1.	Most humans are mortal.	HtM
2.	Most Greeks are humans.	GtH
⊬	All Greeks are mortal.	GaM

Table 5.8: Extension of valid syllogistic moods adapted from (Thompson 1982)

	Figure 1	Figure 2	Figure 3	Figure 4
With "most"	aat att ati ead etd eto	aed add ado ead etd eto	ati eto tai dao	aed eto tai
With "many"	aak atk aki akk eag etg eko ekg	aeg adg ago agg eag etg eko ekg	aki eko kai gao	aeg eko kai
With "few"	aap app apt apk api eab epb epd epg epo	aeb abb abd abg abo eab epb epd epg epo	pai epo bao api	aeb pai epo

5.3 The TFL⁺ Framework: A Tweaked Version of Syllogistic

The plus-minus algebra, as we have seen, provides a simple and logically sound algebraic approach for syllogistic that, alas, does not cover cases of common sense reasoning involving non-classical quantifiers such as "most", "many", or "few"; on the other hand, the syllogistic extended with extra quantifiers comprises a wide range of common sense inference patterns but, unfortunately, it lacks an algebraic procedure. So, given this state of affairs, we produce a tweaked version of syllogistic that includes such a broad range of inferential patterns but with the virtues of an algebraic approach. In order to reach this goal, we proceed in three steps. First we propose a modified syntax of the TFL framework as to include a way to represent the extra quantifiers of the SYLL⁺ framework, then we modify the method of decision of the plus-minus algebra, and finally we show this modification is reliable.

5.3.1 Step 1. The plus-minus algebra meets the extra quantifiers

In order to represent propositions p, t, k, b, d, and g within the framework of the plus-minus algebra, let us consider the proposal displayed in Table 5.9.

Table 5.9: Representation of the syllogistic propositions

Proposition		Representation	Proposition		Representation
SaP	:=	$-S^0+P^0$	SeP	:=	$-S^0-P^0$
SpP	:=	$+S^3+P^0$	SbP	:=	$+S^3-P^0$
StP	:=	$+S^2+P^0$	SdP	:=	$+S^2-P^0$
SkP	:=	$+S^1+P^0$	SgP	:=	$+S^1-P^0$
SiP	:=	$+S^0+P^0$	SoP	:=	$+S^0-P^0$

The rationale behind this proposal is simple: conforming to the SYLL$^+$ framework, non-universal intermediate propositions, namely p (b), t (d), and k (g), are particular to some extent, just like type i (o) propositions, and that requires us to pick, following the TFL framework, a +/+ combination of terms for the positive propositions; and a +/- combination of terms for the negative. However, this is not enough because, according to the SYLL$^+$ framework, propositions p (b), t (d), and k (g) are not convertible,[8] and so, they are not equivalent to propositions of type i (o), which forces us to use some sort of flag in order to explicitly denote this fact: we use superscript indexes.

Now, according to the SYLL$^+$ framework, the new quantifiers imply some sort of order (p (b) implies t (d), t (d) implies k (g), and k (g) implies i (o): *vide* Figure 1) and so the superscript indexes are used not only as flags but also as ordered levels of quantification. This choice has the following features: propositions a, e, i, and o have level 0 to denote the fact that they behave as usual, as if no modifications were made; the superscript indexes are attached to both terms as to specify the detail that propositions p, t, k, b, d, and g are not convertible; and

---

[8] So, for example, t:=*Most Americans speak English* is particular, just like i:=*Some Americans speak English*, but clearly t is not convertible and thus it is not equivalent to i: regard that if *Some Americans speak English* then surely *Some English speakers are American*, but *Most Americans speak English* does not entail *Most English speakers are American*. Similar counter-examples can be developed to show that propositions p (b), t (d), and k (g) do not collapse into i (o) propositions (Cf. Thompson 1986, p. 79).

also, such indexes help us induce an order (3≥2≥1≥0) that indicates that a (e) does not entail p (b), t (d), k (g), i (o); but p (b), t (d), k (g) do entail i (o).[9]

5.3.2 Step 2. The plus-minus algebra modification

Given this new representation, the modification of the plus-minus algebra method of decision is as follows: a conclusion follows validly from a set of premises if and only if *i)* the sum of the premises is algebraically equal to the conclusion, *ii)* the number of conclusions with particular quantity is the same as the number of premises with particular quantity, and *iii)* the level of quantification of the conclusion is less than or equal to the maximum level of quantification of the premises. To exemplify this procedure, let us consider a couple of examples, one valid (Table 5.10), one invalid (Table 5.11).

Table 5.10: att-1

	Proposition	Representation
1.	All H are M.	$-H^0+M^0$
2.	Most G are H.	$+G^2+H^0$
⊢	Most G are M.	$+G^2+M^0$

Table 5.11: tta-1

	Proposition	Representation
1.	Most H are M.	$+H^2+M^0$
2.	Most G are H.	$+G^2+H^0$
⊬	All G are M.	$-G^0+M^0$

The performance of this tweaked version of syllogistic may be better appreciated by considering the trade-off between the complexity of the SYLL⁺ framework and the expressive power of the TFL framework regarding common sense reasoning with non-classical quantifiers. To illustrate this, let us consider

---

[9] This is different from the original presentation by Thompson (1982). Thompson allows universal propositions to entail particular propositions, but our version follows the proposal of Sommers and Englebretsen, and so, we would have to add another rule to the SYLL⁺ framework: 3d. If two premises are universal, the conclusion may not be particular (Cf. Note 10).

some examples (Tables 13–16). As expected, this tweaked version of syllogistic allows the valid inference patterns displayed in Table 5.12.[10]

Table 5.12: Valid syllogistic patterns with extra quantifiers in the TFL⁺ framework

	Figure 1	Figure 2	Figure 3	Figure 4
With "most"	att ati etd eto	add ado etd eto	ati eto tai dao	eto tai
With "many"	atk aki akk etg eko ekg	adg ago agg etg eko ekg	aki eko kai gao	eko kai
With "few"	app apt apk api epb epd epg epo	abb abd abg abo epb epd epg epo	pai epo bao api	pai epo

Table 5.13: An invalid reasoning: kaa-1

	Proposition	Representation
1.	Many homeless are ill.	$+H^3+I^0$
2.	This guy is homeless.	$-g^0+H^0$
⊬	This guy is ill.	$-g^0+I^0$

---

10 For the valid inferential patterns that need existential import, like aat-1 or aak-1, the only requirement is to add the missing implicit premise that states the existence of the minor term, namely, something akin to +S+S: such addition allows the introduction of the valid patterns that appear in Table 5.8 but are absent in Table 5.12.

Table 5.14: An invalid reasoning: akt-4

	Proposition	Representation
1.	All cops are fascists.	$-C^0+F^0$
2.	Many men are cops.	$+M^1+C^0$
⊬	Most men are fascists.	$+M^2+F^0$

Table 5.15: A valid reasoning: bao-3

	Proposition	Representation
1.	Few cars are hybrid.	$+C^3-H^0$
2.	All cars are expensive.	$-C^0+E^0$
⊢	Some expensive cars are not hybrid.	$+E^0-H^0$

Table 5.16: A valid reasoning: ekg-2

	Proposition	Representation
1.	No fool is a citizen.	$-F^0-C^0$
2.	Most voters are citizens.	$+V^2+C^0$
⊢	Many voters are not fools.	$+V^1-F^0$

As we can see from these examples, the TFL⁺ framework gains the advantages of an algebraic method (a reduction of a complex set of rules to a simple and unified formal approach) and, at the same time, it gains the advantages of a theory of syllogisms with non-classical quantifiers (an assessment of a wide range of common sense reasoning patterns that extends the scope of traditional syllogistic).

5.3.3 Step 3. Reliability

Finally, we provide some evidence that this tweaked syllogistic is reliable in so far as the modified TFL⁺ procedure is equivalent to the rules of the SYLL⁺ framework (provided we add the rule 3d, that if two premises are universal, the conclusion may not be particular). In short, for syllogisms at level 0 of quantification, TFL⁺ and SYLL⁺ collapse. For the rest of syllogisms, rules 1a, 1b, 2a, 2b, and 2c of the SYLL⁺ framework are preserved by conditions *i)* and *ii)* of the modi-

5.3 The TFL⁺ Framework: A Tweaked Version of Syllogistic

fied plus-minus algebra method of decision; while rules 3a, 3b, and 3c (and 3d) are preserved by condition *iii)* (and *ii)*).

So, let us say a syllogism is SYLL$^+_{valid}$ (i.e., it is valid in the SYLL$^+$ framework plus rule 3d) if and only if it is TFL$^+_{valid}$ (i.e., it is valid in the modified plus-minus algebra). Consider that, at level 0 of quantification, the proof is trivial: all SYL-L$^+_{valid}$ syllogisms are TFL$^+_{valid}$ and vice versa. But for the remaining syllogisms, regard: if a syllogism is SYLL$^+_{valid}$, then it is TFL$^+_{valid}$. For *reductio*, suppose s is an arbitrary syllogism that is SYLL$^+_{valid}$ but not TFL$^+_{valid}$. Then, s follows the rules of the SYLL$^+$ framework but violates at least one condition of the modified plus-minus algebra. Let us list the conditions that make s a SYLL$^+_{valid}$ syllogism. (*a*) If s follows rule 1a, at least one middle term of s has the minus sign. (*b*) If s follows rule 1b, at least one term in the conclusion has the minus sign or no term in the conclusion has the minus sign. (*c*) If s follows 2a, at least one premise has a predicate-term with a plus sign. (*d*) If s follows rules 2b or 2c, the conclusion and one premise of s must have a predicate-term with a minus sign. (*e*) If s follows 3a, the level of quantification of one premise is 3 and the level of quantification of the conclusion is less than or equal to 3. (*f*) If s follows 3b, the level of quantification of one premise is 2 and the level of the conclusion is less than or equal to 2. (*g*) If s follows 3c, the level of quantification of one premise is 1 and the level of the conclusion is less than or equal to 1. Finally, (*h*) if s follows 3d, the conclusion of s must have a subject-term with a minus sign.

By taking the adequate combinations of the above conditions that rule out the application of (*h*) (because the application of (*h*) is already taken into consideration at the level 0 of quantification), we can construct a set of arbitrary valid syllogisms for any terms X, Y, Z. For example, combination I results from applying the sequence of conditions (*a*), (*b*), (*c*), (*e*), (*f*), and (*g*). The remaining combinations are displayed in Table 5.17.

Table 5.17: Valid syllogistic forms in TFL+ according to the SYLL+ rules

	I		II		III		IV
1.	$-Y^0+Z^0$	1.	$-Y^0-Z^0$	1.	$-Z^0+Y^0$	1.	$-Z^0-Y^0$
2.	$+X^{3,2,1,0}+Y^0$	2.	$+X^{3,2,1,0}+Y^0$	2.	$+X^{3,2,1,0}-Y^0$	2.	$+X^{3,2,1,0}+Y^0$
⊢	$+X^{3,2,1,0}+Z^0$	⊢	$+X^{3,2,1,0}-Z^0$	⊢	$+X^{3,2,1,0}-Z^0$	⊢	$+X^{3,2,1,0}-Z^0$
	V		VI		VII		VIII
1.	$-Y^0+Z^0$	1.	$-Y^0-Z^0$	1.	$+Z^{3,2,1,0}+Y^0$	1.	$-Z^0-Y^0$
2.	$+Y^{3,2,1,0}+X^0$	2.	$+Y^{3,2,1,0}+X^0$	2.	$-Y^0+X^0$	2.	$+Y^{3,2,1,0}+X^0$
⊢	$+X^0+Z^0$	⊢	$+X^0-Z^0$	⊢	$+Z^{3,2,1,0}+X^0$	⊢	$+X^0-Z^0$

Combination I is TFL^+_{valid}: *i)* the sum of the premises takes away the term-schema Y and yields the conclusion with terms +X+Z; *ii)* the number of conclusions with particular quantity is the same as the number of premises with particular quantity (in this case such number is 1); and *iii)* the level of quantification of the conclusion is less than or equal to the maximum level of quantification of the premises: when premise 2 is $+X^3+Y^0$, the conclusion might be $+X^3+Z^0$, or $+X^2+Z^0$, or $+X^1+Z^0$, or $+X^0+Z^0$; when premise 2 is $+X^2+Y^0$, the conclusion might be $+X^2+Z^0$, or $+X^1+Z^0$, or $+X^0+Z^0$; when premise 2 is $+X^1+Y^0$, the conclusion might be $+X^1+Z^0$, or $+X^0+Z^0$; finally, when premise 2 is $+X^0+Y^0$, the TFL framework collapses with the $SYLL^+$ framework and so the conclusion is $+X^0+Z^0$. The remaining combinations also comply with conditions *i)*, *ii)*, and *iii)*.

So, the arbitrary syllogisms defined in Table 5.17 are TFL^+_{valid} as they comply with conditions *i)*, *ii)*, and *iii)* of the modified plus-minus algebra, but s must have the form of one of such syllogisms, since s was constructed by an application of a sequence of conditions that make s a $SYLL^+_{valid}$ syllogism. Hence, s must also be TFL^+_{valid}, but this contradicts the assumption that s is $SYLL^+_{valid}$ but not TFL^+_{valid}. Conversely, from right to left, the outline is more or less direct: if we write down the syllogisms that are TFL^+_{valid}, we will see that they correspond to some syllogism already listed in Table 5.8 (provided we take into account what we have explained in Notes 9 and 10).

This sketch of proof indicates that the TFL^+ framework is reliable in that all valid syllogisms in the $SYLL^+$ framework can be obtained by applying the modified plus-minus algebra method of decision, and vice versa, all syllogisms that can be obtained by applying the modified plus-minus algebra method of decision are valid syllogisms in the $SYLL^+$ framework.

5.4 The $TFL^\oplus$ Framework: A Diagrammatic Extension

Along with TFL, Englebretsen (1992a, 1996) developed a diagrammatic system that represents it. Its syntax requires two diagrammatic objects: the dot and the straight line. In order to represent syllogistic, it provides the syntax displayed in Figure 5.2.

5.4 The TFL⊕ Framework: A Diagrammatic Extension — 177

(A)

(I)

Figure 5.2: Syntax for Englebretsen's Linear Diagrams Adapted From (Englebretsen 1992a)

Given this diagrammatic representation, the system offers a clear decision procedure: a syllogism is valid if and only if the diagram of the conclusion is automatically represented after drawing down the diagram of the premises. So, for example, syllogisms aaa-1 and aii-1 can be represented by the diagrams depicted in Figures 5.3a and 5.3b.

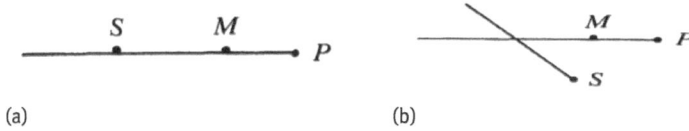

(a) (b)

Figure 5.3: Examples of Syllogisms With Linear Diagrams Adapted From (Englebretsen 1992a)

Englebretsen's diagrams provide a simple and clean diagrammatic approach for syllogistic that, alas, does not cover cases of common sense reasoning involving non-classical quantifiers such as "most", "many", or "few". Now, since we already have sentential systems to deal with the extra quantifiers but we lack the corresponding diagrammatic devices, we produce TFL⊕, a diagrammatic system designed to capture TFL⁺, in three steps. First we propose a modified syntax of Englebretsen's diagrams as to include a way to represent TFL⁺, then we adapt the method of decision of the linear diagrams, and finally we show this modification is reliable.

5.4.1 Step 1. The linear diagrams meet the extra quantifiers

Basically, we propose a vocabulary with the next basic diagrammatic objects: the straight line, the dot, and the circle (Figure 5.4). Clearly, TFL$^\oplus$ is an extension of Englebretsen's diagrams.

Figure 5.4: Vocabulary for TFL$^\oplus$

With the aid of these diagrammatic objects we propose a modification of Englebretsen's linear diagrams in order to represent TFL$^+$ (Figure 5.5).

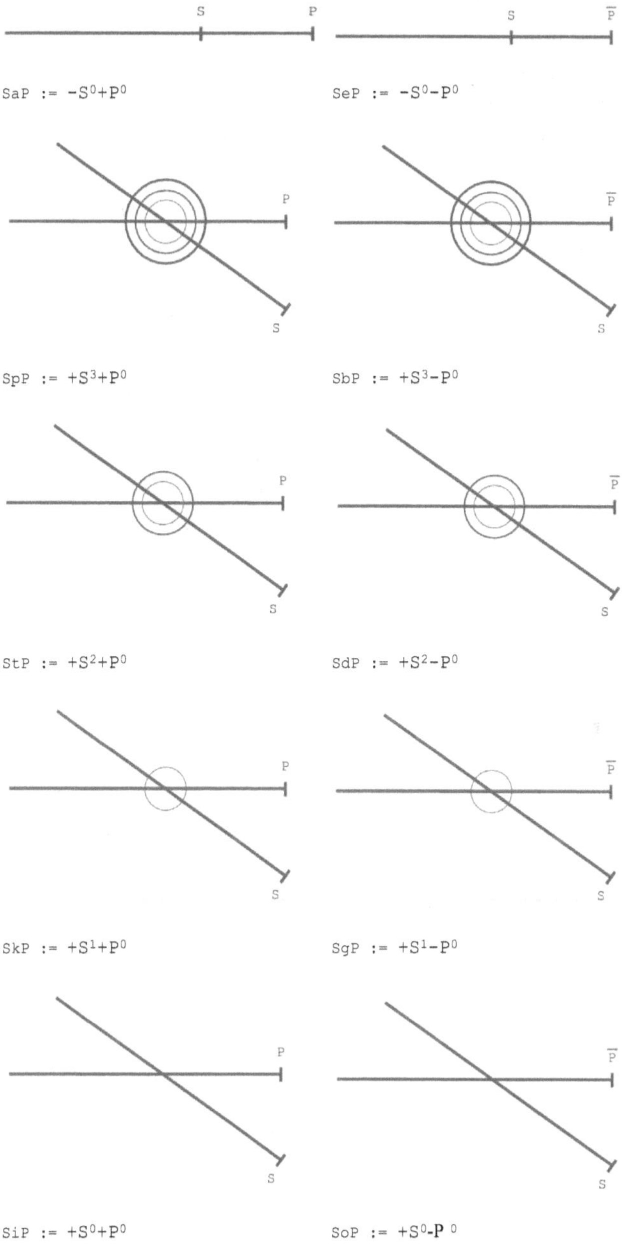

5.4 The TFL$^\oplus$ Framework: A Diagrammatic Extension — 179

SaP := $-S^0+P^0$ SeP := $-S^0-P^0$

SpP := $+S^3+P^0$ SbP := $+S^3-P^0$

StP := $+S^2+P^0$ SdP := $+S^2-P^0$

SkP := $+S^1+P^0$ SgP := $+S^1-P^0$

SiP := $+S^0+P^0$ SoP := $+S^0-P^0$

Figure 5.5: Syntax for TFL$^\oplus$

The reason behind this modification is simple: conforming to the TFL⁺ framework, non-universal intermediate propositions, namely p (b), t (d), and k (g) are particular to some extent but have an index, and so we have to retain Englebretsen's notion of linear intersection but with the addition of circles enclosing the very intersections. This choice has the following features: since propositions a, e, i, and o have level 0 to denote the fact that they behave as usual, so the TFL⊕ diagrams for a, e, i, and o have zero enclosing circles to denote the fact that they behave as usual. However, since TFL⁺ requires the new quantifiers to imply some sort of order (p (b) implies t (d), t (d) implies k (g), and k (g) implies i (o)), the superscript indexes are associated with a certain number of enclosing circles. Plus, since the indexes induce an order (3≥2≥1≥0), the number of circles induce an order (3≥2≥1≥0) that indicates that a (e) does not entail p (b), t (d), k (g), i (o); but p (b), t (d), k (g) do entail i (o). Finally, since the superscript indexes are attached to both terms as to specify the detail that propositions p, t, k, b, d, and g are not convertible, so the enclosing circles are consistent with the emphasis on the scope of the terms affected by the level of quantification (so, for example, the diagram for $+S^3+P^0$ must be read as having three circles around the intersection on S but not on P).

5.4.2 Step 2. The procedure modification

Given this new diagrammatic representation, the adaptation of the decision procedure is trivial: a syllogism is valid if and only if the diagram of the conclusion is automatically represented after drawing down the diagram of the premises. So, for example, in the valid reasoning shown in Table 5.18, the conclusion gets clearly drawn by drawing down the premises as in Figure 5.6; while in an invalid reasoning, the conclusion does not get automatically drawn (Table 5.19, Figure 5.7).

Table 5.18: An example of a valid reasoning: ekg-1

	Proposition	Representation
1.	No citizen is free.	$-C^0-F^0$
2.	Most adults are citizens.	$+A^2+C^0$
⊢	Many adults are not free.	$+A^1-F^0$

5.4 The TFL⊕ Framework: A Diagrammatic Extension — 181

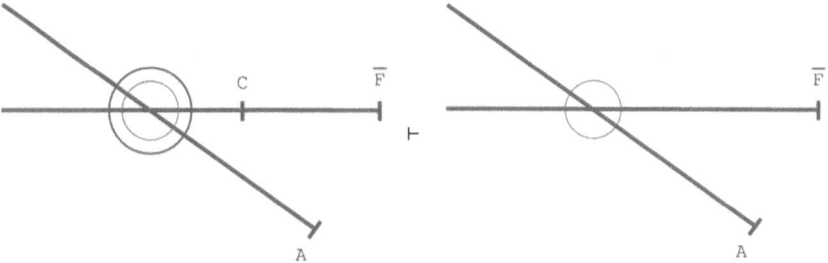

Figure 5.6: A Valid Example: ekg-1 Diagram

Table 5.19: An invalid reasoning: akt-4

	Proposition	Representation
1.	All cops are fascists.	$-C^0+F^0$
2.	Many men are cops.	$+M^1+C^0$
⊬	Most men are fascists.	$+M^2+F^0$

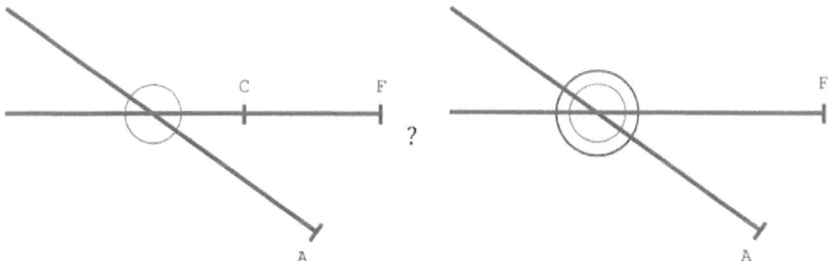

Figure 5.7: An Invalid Example: akt-4 Diagram

As we can see, the TFL⊕ framework gains the advantages of a diagrammatic method (a reduction of an algebraic representation to a simple and unified diagrammatic approach) and, at the same time, it gains the advantages of a theory of syllogisms with non-classical quantifiers (an assessment of a wide range of common sense reasoning patterns that extends the scope of traditional syllogistic). Also, notice that TFL⊕ preserves the capacity to perform inference with relations. Consider, for example, Table 5.20 and the respective diagram in Figure 5.8.

Table 5.20: A valid reasoning with relations

	Proposition	Representation
1.	Few Mexicans are not polite.	$+M^3+P^0$
2.	All polite people give flowers to many girls.	$-P_1^0+(+F^0{}_{12}+G_2{}^1)$
∎	Most Mexicans give flowers to some girls.	$+M_1{}^2+(+F^0{}_{12}+G_2{}^0)$

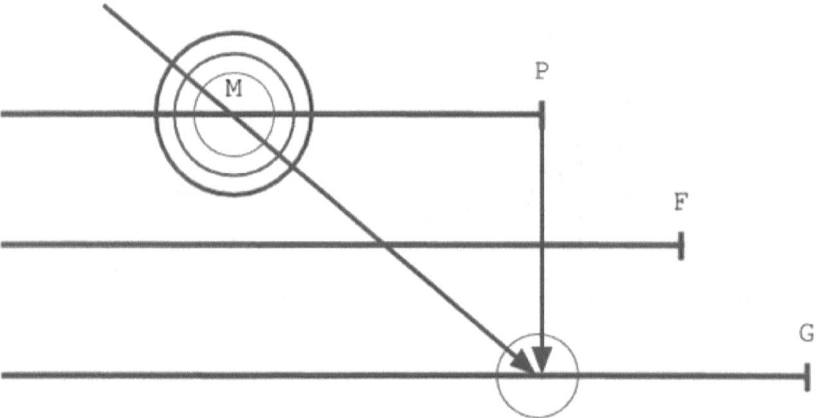

Figure 5.8: A Valid Reasoning With Relations

5.4.3 Step 3. Reliability

As expected, TFL$^\oplus$ is reliable in so far as its results match those of TFL$^+$. Let us say a syllogism is TFL$^+{}_{valid}$ if and only if it is TFL$^\oplus{}_{valid}$. Consider that, at level 0 of quantification, the proof is trivial: all TFL$^+{}_{valid}$ syllogisms are TFL$^\oplus{}_{valid}$ and vice versa, since Englebretsen's diagrams are a straightforward representation of TFL. For the remaining syllogisms, regard: if a syllogism is TFL$^+{}_{valid}$, then its diagram is also TFL$^\oplus{}_{valid}$. For *reductio*, suppose *d* is an arbitrary syllogism that is TFL$^+{}_{valid}$ but is not TFL$^\oplus{}_{valid}$. Then, *d* follows the rules of the TFL framework but violates at least one condition of TFL$^\oplus$. Let us list the conditions that make *d* a TFL$^+{}_{valid}$ syllogism: (*a*') If *d* follows condition *i*), the diagrammatic object of the middle term is drawn in between the diagrammatic objects of the subject and the predicate terms. (*b*') If *d* follows condition *ii*), there must be an intersection of lines to the left of the middle term. (*c*') If *d* follows condition *iii*), the number of circles enclosing a diagrammatic object representing a middle term or a subject term has to follow the order 3≥2≥1≥0. By applying the combination of the above conditions, we can construct a set of arbitrary diagrams for any terms X, Y, Z. For example, combination I' results from directly applying the sequence of conditions (*a'*), (*b'*), and (*c'*). The remaining combinations are displayed in Figure 5.9. Clearly, combination I' is TFL$^\oplus{}_{valid}$: by drawing down the diagrams of the premises we automatically draw the diagram of the conclusion. The remaining combinations behave like so:

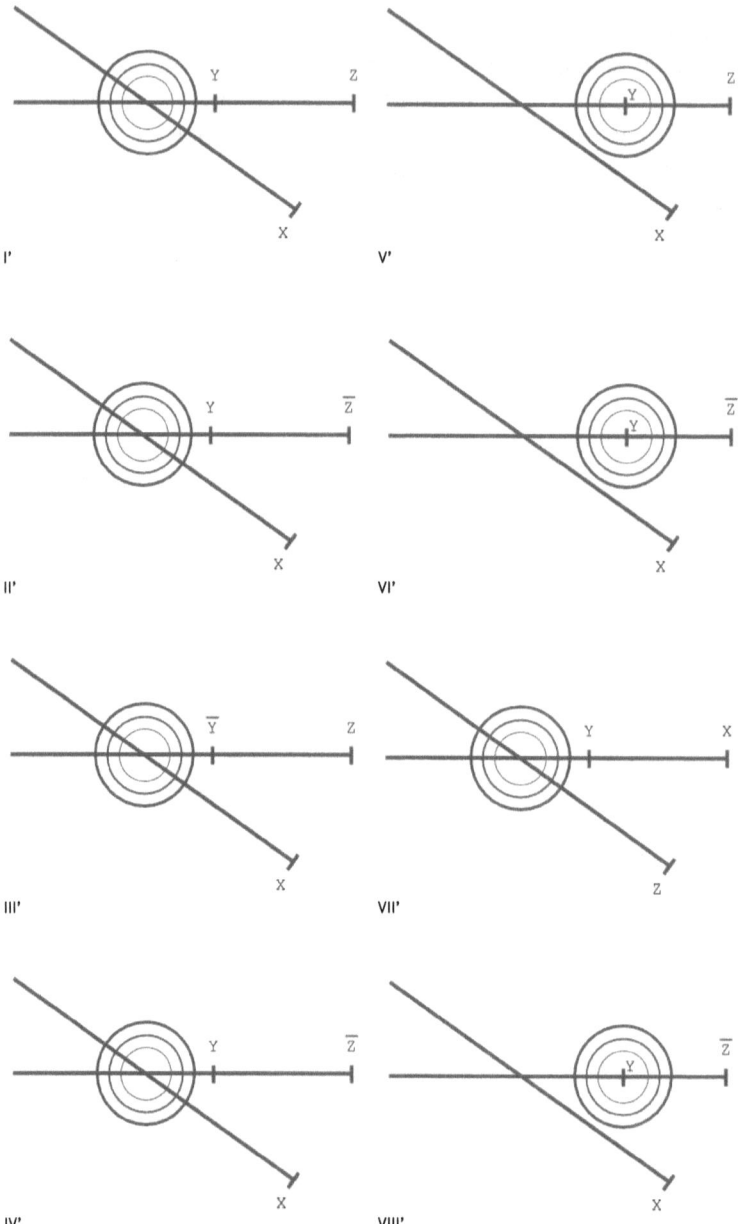

Figure 5.9: Valid Syllogistic Forms in TFL$^\oplus$ According to the TFL$^+$ Conditions

Thus, the arbitrary syllogisms drawn in Figure 5.9 are $\text{TFL}^+_{\text{valid}}$ as they comply with conditions *i)*, *ii)*, and *iii)* of the modified plus-minus algebra, but d must be the diagram of one of such syllogisms, since d was constructed by an application of the sequence of conditions that make d a $\text{TFL}^+_{\text{valid}}$ syllogism. Hence, d must also be $\text{TFL}^\oplus_{\text{valid}}$, but this contradicts the assumption that d is $\text{TFL}^+_{\text{valid}}$ but is not $\text{TFL}^\oplus_{\text{valid}}$. Conversely, from right to left, the outline is also more or less direct: if we draw down the diagrams corresponding to $\text{TFL}^\oplus_{\text{valid}}$ syllogisms, we will see that their sentential counterpart is already listed in Table 5.16. This sketch of proof provides evidence to the fact that the $\text{TFL}^\oplus$ framework is reliable in that all valid syllogisms in the TFL^+ framework can be obtained by drawing down the modified diagrams of $\text{TFL}^\oplus$, and vice versa, all syllogisms that can be drawn by applying the modified diagrams are valid syllogisms in the TFL^+ framework.

5.5 Conclusions

Under direct influence of (Mostowski 1957, Sommers 1967, Peterson 1979, Thompson 1982, Sommers 1982, Englebretsen 1987b), we have presented a reliable extension of syllogistic that deals with a wide range of common sense logical patterns but with the virtues of an algebraic approach (the system TFL^+);[11] and under direct influence of Englebretsen (Englebretsen 1992a, Englebretsen 1996) and TFL^+, we have offered a reliable extension of linear diagrams in order to deal with such a wide range of common sense logical patterns (the system $\text{TFL}^\oplus$).

We believe this set of extensions is quite promising, not only as yet another set of critical thinking tools or didactic contraptions, but as a set of research devices for probabilistic reasoning (in so far as they could be used to represent probabilistic reasoning (Cf. Thompson 1986)), psychology (as they could be used to approximate a richer psychological account of common sense reasoning (Cf. Khemlani & Johnson-Laird 2012)), AI (as they could be used to develop tweaked inferential engines for Aristotelian databases (Cf. Mozes 1989)), and of course, philosophy of logic (as they promote the revision and revival of term logic (Cf. Veatch 1952, Sommers 1982, Englebretsen 1996, Englebretsen

---

[11] Our approach is different, for example, from the one proposed by (Westerståhl 1989) or (Murphree 1998) because, respectively, our version uses term logic *à la* Sommers, and not first-order logic; and vague intermediate quantifiers, and not precise numerical-wise quantifiers.

and Sayward 2011) and logic diagrams as tools that might be more interesting and powerful than once they seemed (Cf. Carnap 1930, Geach 1972)).

Appendix: Rules of inference for TFL

For the purposes of this study, here we expound the rules of inference for TFL as they appear in (Englebretsen 1996).
1. Rules of immediate inference.
    1. Premise (P): Any premise or tautology can be entered as a line in proof. (Tautologies that repeat the corresponding conditional of the inference are excluded. The corresponding conditional of an inference is simply a conditional sentence whose antecedent is the conjunction of the premises and whose consequent is the conclusion.)
    2. Double Negation (DN): Pairs of unary minuses can be added or deleted from a formula (i.e., $--X=X$).
    3. External Negation (EN): An external unary minus can be distributed into or out of any phrase (i.e., $-(\pm X \pm Y) = \mp X \pm Y$).
    4. Internal Negation (IN): A negative qualifier can be distributed into or out of any predicate-term (i.e., $\pm X \cdot (\pm Y) = \pm X + (\pm Y)$).
    5. Commutation (Com): The binary plus is symmetric (i.e., $+X+Y=+Y+X$).
    6. Association (Assoc): The binary plus is associative (i.e., $+X+(+Y+Z)=+(+X+Y)+Z$).
    7. Contraposition (Contrap): The subject- and predicate-terms of a universal affirmation can be negated and can exchange places (i.e., $-X+Y=-(-Y)+(-X)$).
    8. Predicate Distribution (PD): A universal subject can be distributed into or out of a conjunctive predicate (i.e., $-X+(+Y+Z)=+(-X+Y)+(-X+Z)$) and a particular subject can be distributed into or out of a disjunctive predicate (i.e., $+X + (-(-Y) - (-Z)) = -(+X + Y) - (+X + Z)$).
    9. Iteration (It): The conjunction of any term with itself is equivalent to that term (i.e., $+X+X=X$).
2. Rules of mediate inference.
    1. (DON): If a term, M, occurs universally quantified in a formula and either M occurs not universally quantified or its logical contrary occurs universally quantified in another formula, deduce a new formula that is exactly like the second except that M has been replaced at least once by the first formula minus its universally quantified M.
    2. Simplification (Simp): Either conjunct can be deduced from a conjunctive formula; from a particularly quantified formula with a conjunctive

subject-term, deduce either the statement form of the subject-term or a new statement just like the original but without one of the conjuncts of the subject-term (i.e., from +(+X+Y)±Z deduce any of the following: +X+Y, +X±Z, or +Y±Z), and from a universally quantified formula with a conjunctive predicate-term deduce a new statement just like the original but without one of the conjuncts of the predicate-term (i.e., from -X ±(+Y+Z) deduce either -X±Y or -X±Z).

3. Addition (Add): Any two previous formulae in a sequence can be conjoined to yield a new formula, and from any pair of previous formulae that are both universal affirmations and share a common subject-term a new formula can be derived that is a universal affirmation, has the subject-term of the previous formulae, and has the conjunction of the predicate-terms of the previous formulae as its predicate-term (i.e., from -X+Y and -X+Z deduce -X+(+Y+Z>).

6 Drawing Conclusions: Get Better Diagrams

Tell me and I will listen. Show me and I will understand.

Lakota proverb

"... what is the use of a book," thought Alice, "without pictures or conversations?"

Lewis Carroll

What's the use of a book without diagrams or equations?

Melina Sydney Padua

I never read. I just look at pictures.

Andy Warhol

As the above hasty survey must have made evident, there are innumerable unsolved problems in the subject, and much work needs to be done.

Russell

There have always been logicians and mathematicians who have been ready to argue that diagrams might well be acceptable heuristic aids in teaching, and even engaging in, the process of logical deduction, but, they continue, diagrams cannot possibly replace the standard practice of carrying out such tasks in the medium of a symbolic, non-graphic system. The usual argument for this rests on the conviction that diagrams can (implicitly or explicitly) import illegitimate assumptions that are simply due to the singular nature of the diagram, and this is the source of potential error. So, how could one possibly avoid errors endemic to the diagrams we use? Catherine Legg has a fit response: "The short answer is: Get better diagrams ('better' here meaning more perspicuous)" (Legg 2013, 16). Of course, there is a further response, which is simply revealed by the examination of the many results of ongoing work aimed at improving, extending, strengthening (or even replacing) the systems of Euler, Venn, Carroll, Peirce, etc. A system of sufficiently perspicuous diagrams is an efficient, practical, and effective tool for modelling and carrying out our tasks of ratiocination (e.g., detecting deductive validity or invalidity, determining consistency or inconsistency of suites of information, deduction). Such diagrams provide the free-rides that ground our perceptual intuitions, intuitions that are of little import when using only purely symbolic, linguistic, formal systems.

> What ... makes us persist in seeing diagrammatic proofs as somehow 'more intuitive' than language ones? The main factor is that the perceptual elements utilized in diagrammatic proofs are *obviously* geometric in character; although exactly the same is true of language elements, we're not aware of that. (Azzouni 2013, 333)

These "perceptual elements" are primarily the objects of visual perception. Consider how often we resort to metaphorical language when talking about certain cognitive activities (conceiving, considering, attending, understanding, knowing, etc.). Sometimes we reach for the tactile metaphor ('grasp', 'touch', 'put a finger on', 'wrap my arms/mind around', etc.). However, when our talk touches on the more particularly epistemic (knowing, understanding, learning, informing, etc.), we turn to the visual ('see', 'show', 'reveal', 'demonstrate', etc.).

A system of better logic diagrams, more perspicuous diagrams, is of little use if it cannot provide for diagrams that are adequately suited for representing and for carrying out the usual tasks assigned to deductive logic; in other words, such a system must be both expressively and inferentially powerful enough to be useful as a practical tool. They are, in effect, tools of reason, they "provide the means for us to exercise our rational intuitions" (Legg 2013, 16). In fact, it can be (and has been) argued that, more particularly, diagrams can be understood as scientific instruments for carrying out tasks in logic and mathematics (see Moktefi 2017). "We do not make derivations with diagrams. We rather use imagination, not rules, to manipulate diagrams in order to solve problems. ... we work *with* diagrams (as tools), rather than *on* diagrams (as objects)" (Moktefi 2017, 82). The role of any scientific instrument (e. g., a microscope, CERN's Large Hadron Collider, a Euclidian diagram) is determined by the ways it is used in practice by scientists. This is the *practical* (i.e., practice-based) view of logic diagrams. In a particularly clear and insightful defense of this view, Valeria Giardino has contrasted it with both the "suspicious" view and the "tame" view (Giardino 2013, 136–137). The former is the usual view that diagrams used in mathematics or logic might be useful in the process of discovery but not in the process of justification (proof). Consequently, on such a view, the reliance on diagrams here is suspicious. By contrast, according to the tame view, "proofs are at the core of logic and mathematics and ... proofs are derivations" (Giardino 2013, 137). The appropriate strategy, then, according to such a view is that diagrammatic reasoning has to be "tamed" by subjecting diagrammatic manipulations to syntactic rules so that they "can become 'rigorous' elements of formal proofs" (Giardino 2013, 137). However, diagrams *do not work* like sentences" (Giardino 2013, 138) and "formal proofs are only a proper subset of a variety of justifications" (Giardino 2013, 140).

The practical (practice-based) view, according to Giardino, provides "an account of reasoning *in* not *from* diagrams based on the conception of diagrams as tools used within a specific practice. ... diagrams are not pictures of abstract objects but tools for reasoning about abstract relationships" (Giardino 2013, 139). Diagrams are "cognitive and epistemic tools" and "must always be *interpreted* within a certain *context of use*." They "are given *with an intention*." In short, they "are representations used with the intention of embodying relations; moreover, they promote inference because they can be interpreted and manipulated in various ways according to the shared practice" of the communities of mathematicians and logicians. (Giardino 2013, 142). Giardino also remarks on just how the use of a diagram is intimately coupled with its accompanying linguistic text.

> [T]he text and its accompanying figure are engaged each time in a fruitful and rational intercourse: the text gives the instructions to outline the figure that stands by its side, and to draw the conclusion from it. The verbal instructions do not assume the complete figure at the start, but rather walks the readers through its construction, and therefore the readers go back and forth between verbal and visual. In following these instructions, they *actively imagine* drawing lines.
>
> Instead of accepting the unuseful opposition between visual perception and linguistic knowledge, I propose to focus on this kind of 'manipulative' imagination at play when reasoning with diagrams. (Giardino 2013, 145)
>
> My proposal is that in diagrammatic reasoning what counts is not the appearance of a diagram and a list of explicit rules that can be applied to it, but rather a set of *procedures:* when one learns to use a certain diagrammatic system for performing some inferences, she learns a *manipulation practice*. The diagram becomes the mathematician's worksite, where operations, plans, and experiments are made in order to find solutions and reasons for these solutions. While syntactic rules are piecemeal, procedures are holistic. (Giardino 2013, 145–146)

The advantage of the practice-based view is that, in contrast with the "suspicious" view and the "tame" view, it provides a clear and accurate account of how diagrams are *actually used* by mathematicians and logicians when they perform their rational tasks. In spite of the fact that all human performance (including by mathematicians and logicians) is subject to error, "the diagram does not need to be properly drawn, as long as the user is aware of the prescriptions contained in the instructions for its construction and is aware of its intended meaning" (Giardino 2013, 149). The idea that the diagrams used by mathematicians and logicians are practical tools used to carry out real tasks of reasoning, not just instructional or heuristic devices, and not just alternative symbolizations, is a good idea that comports well with empirical facts. Diagrams *are* the mathematician's and logician's instruments (as well as worksites). This does not mean that diagrammation is inherently simple or easy.

> Once one has been shown the diagram, and has been shown how to use it to make the passage from the starting point to the endpoint, one can see how the diagram serves to establish the result, and one can reproduce the result. The hard part, the part that can take real mathematical genius, is finding the diagram." (Macbeth 2012b, 72)

This is not only the case for the mathematician but for the logician as well. But how *does* the logician find the right diagram? Moktefi has provided an indication of how to begin to find the answer: "The moral of the story is that when we reason with logic diagrams, we do not merely search for the best diagrams to *reason* with, we also look for the best logic to diagram with" (Moktefi 2015, 612).

On reflection, this has been the case with the development of the ED system of logic diagrams. I had come to understand, accept, and then help contribute to the development and deployment of Sommers' Term Functor Logic (TFL). Eventually, I hoped to couple TFL with an appropriate system of diagrams. In effect, I simply puttered around in my "worksite" hoping to find the simplest, most useful array of graphic tools for dealing with basic deductions (initially, nothing more than classical syllogisms). Good tools are generally ones that are relatively easy to use, accessible to most people who need them, and efficient for use in dealing with the appropriate tasks (and, in some cases, can be used for secondary tasks). Usually, when one needs a particular tool, the best course is to obtain one from an appropriate dealer. I tried various stores: Euler's, Venn's, Peirce's, Euler's again, and so on. Next, I turned to considering using line segments rather than closed figures. However, the systems that had been devised by Leibniz and then Lambert could not, on their own, be adequate. Their diagrams did not accommodate representations of singulars, relationals, or negative terms. As well, they could not even display the symmetry of either E or I categoricals. I finally decided to let TFL be my direct guide here. This meant that what was required was a system of diagrams that treated all terms as logically "charged" (either positive or negative), incorporated singular terms as well, and analyzed both relational terms and entire sentences as nothing more than syntactically complex terms (copulated term pairs).

Though nothing is perfect, there are good tools, very good tools, and not so good tools. The test is whether or not they can be used to get the job done. Many systems of logic diagrams have been challenged by their limited abilities to deal with negative expressions, relational terms, too many terms, overdetermined alternatives, etc. (see Gurr, Lee, and Stenning 1998). ED certainly has its limitations. For example, as we have seen, it suffers, as two dimensional graphic systems generally do, from an inability to go beyond the geometric bounds set by Helly's Theorem. Nevertheless, such systems are empowered in various ways by other geometric constraints, such as those that can license "free rides" for inferences. My conviction is that ED is powerful enough (both expressively and inferentially), simple enough, uncluttered enough, and accessible enough to be considered for use as an instrument of reasoning. As we have seen, it is, to a large degree, cognitively efficient. Moreover, it is flexible enough to offer, in many cases, more than one choice of diagrammatic representation (e. g., Simple versus Full diagrams). Given the importance of the "fruitful and rational intercourse" (as Giardino called it) between constructing, interpreting, and manipulating by use of one's active imagination and the logic diagram it accompanies, the language of that text is crucial. Since most reasoning is not carried out in the medium of a formalized language, that text, if it is to be formalized, should be one that cleaves as closely as possible to natural language. That's just what TFL aims to do. So, ED is a good candidate for reflecting how we ordinarily carry out tasks of ratiocination, while at the same time serving as a powerful tool for the tasks required of those who constitute the community of logicians.

It seems safe to say that today millions of people are familiar with Venn diagrams. The system is often introduced in school; it is used by many people as a tool to illustrate various set relations involved in mathematics, economics, military strategy, epidemiology, etc. Venn diagrams are useful, perspicuous, and relatively simple (at least when only a small number of sets is involved). Yet, though Venn meant his diagrammatic system to be a graphic version of Boole's algebra, few people familiar with his diagrams have any significant knowledge of, or interest in, the algebra. Venn's system can be considered (and used) on its own, independently of Boole's algebra. In the same way, it is possible for ED to be used on its own, ultimately independently of TFL. Such line diagrams are useful, perspicuous, and relatively simple. In comparing various diagram systems, including linear (though more like Leibniz-Lambert diagrams with overlapping parallel line segments of different sizes rather than line segment inclusion or intersection), empirical studies have shown that linear diagrams are relatively less prone to clutter and are more readily used by subjects. "Our results suggest that linear diagrams should be more widely adapted, at least for use by the general population" (Chapman, Stapleton, Rogers, Miallef, and Blake 2014, 59; also Rogers, Stapleton, and Chapman 2015). Having introduced ED to many students over a number of years, I came to see that these diagrams are easier to construct than most other types of logic diagrams – easier digitally, easier for quick, paper-and-pencil, back-of-the-envelope logical calculation, probably easier even with a stick on the dry ground of the Lyceum.

A final thought: It would appear that we are now living in the best time to be interested in logical diagramming – a new Golden Age. A growing number of philosophers, logicians, mathematicians, linguists, cognitivist psychologists, and knowledge engineers are expanding existing systems and developing new diagrammatic ones. They are addressing the questions of the efficacy, range, limits, and expressiveness of these systems. They are attempting to account for how such systems can model, or even facilitate actual human reasoning. The literature devoted to these issues just in the past thirty years is large and expanding (see the Bibliography and Further Readings for a sampling). My hope is that this book makes a small contribution to the overall logic diagram project. In any event, there is much to look forward to from this community of scholars. The best is yet to come.

Bibliography and Further Reading

> We call on others to aid us in deliberating on important questions – distrusting ourselves as not being equal to deciding.
>
> Aristotle

> Paradise will be a kind of library.
>
> Borges

Abeles, F., 2007, "Lewis Carroll's Visual Logic," *History and Philosophy of Logic*, 28: 1–17.
Abeles, F., 2010, *The Logical Pamphlets of Charles Lutwidge Dodgson and Related Pieces*, New York: The Lewis Carroll Society of North America.
Ackrill, J.L. (ed), 1987, *A New Aristotle Reader*, Princeton, NJ: Princeton University Press.
Allwein, G. and Barwise, J. (eds), 1996, *Logical Reasoning with Diagrams*, New York: Oxford University Press.
Alvarez, E. and Correia, M., 2012, "Syllogistic with Indefinite Terms," *History and Philosophy of Logic*, 4: 297–306.
Aristotle, 1941, *Metaphysics*, W.D. Ross (transl), in: *The Basic Works of Aristotle*, R. McKeon (ed), New York: Random House.
Aristotle, 1949, *Aristotle's Prior and Posterior Analytics*. W.D. Ross (ed), Oxford: Oxford University Press.
Aristotle, 1960, *Posterior Analytics*, H. Tredennick (transl), *Topica* E.S. Forster (transl), Cambridge, MA: Harvard University Press, Loeb Classic Library.
Aristotle, 1962, *Prior Analytics*, H. Tredennick (transl), *The Categories* and *On Interpretation*, H.P. Cook (transl), Cambridge, MA: Harvard University Press, Loeb Classic Library.
Aristotle, 1963, *Categories and De Interpretatione*, J.L. Ackrill (transl), Oxford: Clarendon Press.
Aristotle, 1987, *Prior Analytics*, P. Geach (transl), in: *A New Aristotle Reader*, J. Ackrill (ed), Princeton, NJ: Princeton University Press.
Aristotle, 1989, *Prior Analytics*, R. Smith (transl), Indianapolis: Hackett.
Aristotle, 1995, *Aristotle: Selections*, T. Erwin and G. Fine (eds), Indianapolis: Hackett.
Arkoudas, K. and Brinsjord, S., 2009, "Vivid: A Framework for Heterogeneous Problem Solving," *Artificial Intelligence*, 173: 1367–1405.
Arnauld, A. and Nicole, P., 1964 [1662], *The Art of Thinking: The Port Royal Logic*, J. Dickoff and P. James (transl), NY: Bobbs-Merrill.
Azzouni, J., 2013, "That We *See* That Some Proofs Are Perfectly Rigorous," *Philosophia Mathematica*, 21: 323–338.
Bacon, J. 1985, "The Completeness of a Predicate-Functor Logic," *Journal of Symbolic Logic*. 50: 9093–926.
Bacon, J., 2000, "Syllogistica Carolina Rediviva," *Uppsala Prints and Preprints in Philosophy*, 2: 1–26.
Barker-Plummer, D., Beaver, D., van Benthem, J., and di Luzio, P.S. (eds), 2002, *Words, Proofs, and Diagrams*, Stanford: CSLI Publications.
Baron, M., 1969, "A Note on the Historical Development of Logic Diagrams: Leibniz, Euler and Venn," *The Mathematical Gazette*, 53: 113–125.

Barwise, J. and Etchemendy, J., 1993, *Hyperproof: Logical Reasoning with Diagrams*, Stanford: CSLI Publications.
Barwise, J. and Etchemendy, J., 1995, "Heterogeneous Logic," in: *Diagrammatic Reasoning: Cognitive and Computational Perspectives*, B. Chandrasekaran, J. Glasgow and N.H. Narayanan (eds), Cambridge, MA: MIT Press.
Barwise, J. and Etchemendy, J., 1996, "Visual Information and Valid Reasoning," Logical Reasoning with Diagrams, G. Allwein and J. Barwise (eds), New York: Oxford University Press. Originally in: *Visualization in Teaching and Learning Mathematics*, W. Zimmerman and S. Cunningham (eds), Washington, DC: Mathematical Association of America, 1991.
Barwise, J. and Etchemendy, J., 2002, *Language, Proof, and Logic*, Stanford, CA: CSLI Publications.
Bauer, M. and Johnson-Laird, P.N., 1993, "How Diagrams Can Improve Reasoning," *Psychological Science*, 4: 372–378.
Bellucci, F., 2013, "Diagrammatic Reasoning: Some Notes on Charles S. Peirce and Friedrich A. Lange," *History and Philosophy of Logic*, 34: 293–305.
Bellucci, F., Chiffi, D., and Pietarinen, A.-V., to appear, "Assertive Graphs," *Journal of Applied Non-Classical Logic*,
Bellucci, F., Moktefi, A., and Pietarinen, A.-V., 2014, "Diagrammatic Autarchy: Linear Diagrams in the 17th and 18th Centuries," in: *Diagrams, Logic and Cognition*, J. Burton and L. Choudhury (eds), *CEUR Workshop Proceeding*, 1132: 23–30.
Bellucci, F. and Pietarinen, A.-V., 2016, "Existential Graphs as an Instrument of Logical Analysis: Part I. Alpha," *The Review of Symbolic Logic*, 9: 209–237.
Bellucci, F. and Pietarinen, A.-V., to appear, "From Mitchell to Carus: Fourteen Years of Logical Graphs in the Making," *Transactions of the Charles S. Peirce Society*.
Bennett, D., 2015, "Drawing Logical Conclusions," *Math Horizons*, 22: 12–15.
Bernhard, P., 2001, Euler-Diagramme: *Zur Morphologie einer Repräsentationsform in er Logik*, Paderborn: Mentis.
Bernhard, P., 2012, "The Remarkable Diagrams of Johann Maass," in: *Mathematik – Logik – Philosophie: Ideen und ihre historischen Wechselwirkungen*, G. Löfflat (ed), Frankfurt am Main: Harri Deutsch.
Bird, O., 1964, *Syllogistic and its Extensions*, Englewood Cliffs, NJ: Prentice-Hall.
Blackwell, A.F. (ed), 2001, *Thinking with Diagrams*, Dordrecht: Kluwer.
Blake, A., Stapleton, G., Rogers, P., and Howse, J., 2016, "The Impact of Topological and Graphic Choices on the Perception of Euler Diagrams," *Information Sciences*, 330: 455–482.
Block, N., 1983, "Mental Pictures and Cognitive Science," *The Philosophical Review*, 92: 499–541.
Bocheński, I.M., 1951, "Non-Analytical Laws and Rules in Aristotle," *Methodos*, 3: 70–80.
Bocheński, I.M., 1968, *Ancient Formal Logic*, Amsterdam: North-Holland Publ.
Boole, G., 1854, *The Laws of Thought*, Cambridge, Macmillan; reprinted with an introduction by J. Corcoran, Buffalo: Prometheus Books, 2003.
Boulet, G., Francavilla, M., and Englebretsen, G., 1994, "L'utilisation de modèles linéares dans l'enseignement des entiers relatifs," *Instantanes Mathematiques*, 30: 16–18.
Brady, G., 1997, "From the Algebra of Relations to the Logic of Quantifiers," in: *Studies in the Logic of Charles Sanders Peirce*, N. Houser, N., D.D. Roberts, and J. Van Evra. (eds), Bloomington and Indianapolis: Indiana University Press.

Braine, M.D.S., 1978, "On the Relationship Between the Natural Logic of Reasoning and Standard Logic," *Psychological Review*, 85: 1–21.

Braine, M.D.S. and O'Brien, D.P., 1998, *Mental Logic*, Mahwah, NJ: Erlbaum.

Bullynck, M., 2013, "Erhard Weigel's Contributions to the Formation of Symbolic Logic," *History and Philosophy of Logic*, 34: 25–34.

Burch, R.W., 1997, "Peirce on the Application of Relations to Relations," in: *Studies in the Logic of Charles Sanders Peirce*, N. Houser, N., D.D. Roberts, and J. Van Evra. (eds), Bloomington and Indianapolis: Indiana University Press.

Carey, S., 2009, "Where our Number Concepts Come From," *Journal of Philosophy*, 106: 220–254.

Carnap, R., 1930, "*Die alte und die neue Logik*," *Erkenntnis*, 1: 12–26.

Carroll, L. (Charles Dodgson), 1886, *The Game of Logic*, London: Macmillan.

Carroll, L., 1897, *Symbolic Logic: Part I*, London: Macmillan.

Carroll, L., 1886, *The Game of Logic*, London: Macmillan.

Carroll, L., 1958, *Symbolic Logic* and *The Game of Logic*, New York: Dover.

Carroll, L., 1986, *Symbolic Logic: Part I, Elementary and Part II, Advanced*, W.W. Bartley (ed), New York: Clarkson N. Potter.

Castañeda, H.-N., 1976, "Leibniz's Syllogitico-Propositional Calculus," *Notre Dame Journal of Formal Logic*, 17: 481–500.

Castañeda, H.-N., 1982, "Leibniz and Plato's *Phaedo* Theory of Relations and Predication," in: *Leibniz: Critical and Interpretive Essays*, M. Hooker (ed), Minneapolis: University of Minnesota Press.

Castañeda, H.-N., 1990, "Leibniz's Complete Propositional Logic," *Topoi*, 9: 15–28.

Castro-Manzano, J.M., 2015, "Słupecki's Rule for Diagrammatic Reasoning," *Studia Methodologiczne*, 35: 79–96.

Castro-Manzano, J.M., 2017a, "Re(dis)covering Leibniz's Diagramatic Logic," *Tópicos, Revista de Filosofía*, 52: 89–116.

Castro-Manzano, J.M., 2017b, "Remarks on the Idea of Non-monotonic (Diagrammatic) Inference," *Open Insight*, 8: 243–263.

Castro-Manzano, J.M. and Pacheco-Montes, J.R., 2018, "Moded Diagrams for Moded Syllogisms," *Diagramatic Representation and Inference: 10th International Conference, Diagrams 2018*, Springer, pp. 757–761, online: https://doi.org/10.1007/978-3-319=91376-6. Accessed December 17, 2017.

Castro-Manzano, J.M. and Pacheco-Montes, J.R., to appear, "Syllogistic with Jigsaw Puzzles." *Journal of Logic, Language and Information*.

Champagne, M., 2016, "Brandom, Peirce, and the Overlooked Friction of Contrapiction," *Synthese*, 193: 2561–2576.

Chandrasekaran, B., Glasgow, J., and Narayanan, N., (eds), 1995, *Diagrammatic Reasoning: Cognitive and Computational Perspectives*, Cambridge, MA: AAAI Press/MIT Press.

Chapman, P., Stapleton, G. and Rogers, P., 2014, "PaL: A Linear Diagram-Based Visual Language," *Journal of Visual Language and Computing*, 25: 945–954.

Chapman, P., Stapleton, G., Rogers, P., Miallef, L., and Blake, A., 2014, "Visualizing Sets: An Empirical Comparison of Diagram Types," in: *Diagrammatic Representation and Inference*, T. Dwyer, H. Purchase, and A. Delaney (eds), Berlin: Springer, pp. 146–160.

Chatti, S. and Schang, F., 2013, "The Cube, the Square and the Problem of Existential Import," *History and Philosophy of Logic*, 34: 101–132.

Cheng, P. C.-H., 2014, "Graphical Notations for Syllogisms: How Alternative Representations Impact the Accessibility of Concepts," *Journal of Visual Language and Computing*, 25: 170–185. online: http://dx.doi.org/10.1016/j.jvlc.2013.08.008. Accessed December 19, 2017.

Cheng, P., Lowe, R.K, and Scaife, M., 2001, "Cognitive Science Approaches to Understanding Diagrammatic Representations," *Artificial Intelligence Review*, 15: 79–84.

Chomsky, N., 1980, *Rules and Representations*, Oxford: Blackwell.

Chomsky, N., 1981, *Lectures on Government and Binding*, Dordrecht: Foris.

Choudhury, L., and Chakraborty, M.K., 2004, "On Extending Venn Diagrams by Augmenting Names of Individuals," 3rd *International Conference on the theory and Application of Diagrams*. NY: Springer, pp. 142–146.

Choudhury, L., and Chakraborty, M.K., 2016, "Singular Propositions, Negation and the Square of Opposition," *Logica Universalis*, 10: 215–231.

Church, A., 1956, *An Introduction to Mathematical Logic*, Princeton: Princeton University Press.

Clark, B., 1981, "A Calculus of Individuals Based on 'Connection'," *Notre Dame Journal of Formal Logic*, 22: 204–218.

Clark, B., 1985, "Individuals and Points," *Notre Dame Journal of Formal Logic*, 26: 61–75.

Clark, D.S., 1983, "Negating the Subject," *Philosophical Studies*, 43: 349–353,

Cooper, W.S., 2001, *The Evolution of Logic: Logic as a Branch of Biology*, Cambridge: Cambridge University Press.

Corcoran, J., 1972, "Completeness of an Ancient Logic," *Journal of Symbolic Logic*, 37: 696–702.

Corcoran, J., 1974, "Aristotle's Natural Deduction System," in: *Aristotle's Logic and its Modern Interpretations*, J. Corcoran (ed), Dordrecht: Kluwer.

Corcoran, J., 1999a, "Logical Form," in: *Cambridge Dictionary of Philosophy*, R. Audi (ed), Cambridge: Cambridge University Press.

Corcoran, J., 1999b, "Universe of Discourse," in: Cambridge Dictionary of Philosophy, R. Audi (ed), Cambridge: Cambridge University Press.

Corcoran, J., 2003, "Aristotle's *Prior Analytics* and Boole's *Laws of Thought*, History and Philosophy of Logic, 24: 261–288.

Corcoran, J., 2004, "The Principle of Wholistic Reference," *Manuscrito*, 27: 159–171.

Corcoran, J. 2008, "Meanings of Form," *Manuscrito*, 31: 223–266.

Corcoran, J., 2009, "Aristotle's Demonstrative Logic," *History and Philosophy of Logic*, 30: 1–20.

Corcoran, J. and Scanlan, M., 1982, "The Contemporary Relevance of Ancient Logical Theory," *Philosophical Quarterly*, 32: 76–86.

Corcoran, J. and Wood, S., "Boole's Criteria for Validity and Invalidity," *Notre Dame Journal of Formal Logic*, 21: 609–638.

Crain, S. and Khlentzos, D., 2008. "Is Logic Innate?" *Biolinguistics*, 2: 24–65.

Crain, S. and Khlentzos, D., 2010, "The Logic Instinct," *Mind & Language*, 25: 30–65.

Danka, I., 2016, "Which Way to Argue (For): The Visual, the Linguistic and the Symbolic," *Opus et Educatio*, 3: 151–156.

Dejnožka, J., 2010, "The Concept of Relevance and the Logic Diagram Tradition," *Logica Universalis*, 4: 67–135.

De Morgan, A., 1851, "On the Symbols of Logic, the Theory of the Syllogism, and in particular of the Copula, and the application of the Theory of Probabilities to some questions of evidence," *Transactions of the Cambridge Philosophical Society*, 9: 79–127.

De Morgan, A., 1864, "On the Syllogism IV and on the Logic of Relations," *Cambridge Philosophical Transactions*, 10: 331–338.

De Morgan, A., 1966, *On the Syllogism and Other Logical Writings*, P. Heath (ed), London: Routledge & Kegan Paul.

Dipert, R., 1981, "Peirce's Propositional Logic," *Review of Metaphysics*, 34: 569–595.

Dipert, R., 1995, "Peirce's Underestimated Place in the History of Logic: A Response to Quine," in: *Peirce and Contemporary Thought: Philosophical Inquiries*, K.L. Ketner (ed), New York: Fordham University Press, 32–58.

Dove, I., 2002, "Can Pictures Prove?" *Logique et Analyse*, 179–180: 309–340

Edwards, A.W.F., 2004, *Cogwheels of the Mind: The Story of Venn Diagrams*, Baltimore: Johns Hopkins University Press.

van Eijck, J., 2005, "Natural Logic for Natural Language," in: Logic, Language, and Computation: 6th International Tbilisi Symposium, B. ten Cate and H. Zeevat (eds), Berlin and Heidelberg: Springer, pp. 216–230.

Einarson, B., 1936, "On Certain Mathematical Terms in Aristotle's Logic," *American Journal of Philology*, 57: 33–54 and 151–172.

Englebretsen, G., 1979, "Notes on a New Syllogistic," *Logique et Analyse*, 85–86: 111–120.

Englebretsen, G., 1980a, "On Propositional Form," *Notre Dame Journal of Formal Logic*, 21: 101–120.

Englebretsen, G., 1980b, "Singular Terms and the Syllogistic," *The New Scholasticism*, 54: 68–74.

Englebretsen, G., 1980c, "Noncategorical Syllogisms in the *Analytics*," *Notre Dame Journal of Formal Logic*, 21: 602–608.

Englebretsen, G., 1981, *Three Logicians: Aristotle, Leibniz and Sommers, and the Syllogistic*, Assen: Van Gorcum.

Englebretsen, G., 1982a, "Do We Need Relative Identity?" *Notre Dame Journal of Formal Logic*, 23: 91–93.

Englebretsen, G., 1982b, "Aristotle and Quine on the Basic Combination," *The New Scholasticism*, 56: 244–249.

Englebretsen, G., 1982c, "Leibniz on Logical Syntax," *Studia Leibnitiana*, 14: 119–126.

Englebretsen, G. 1982d, "Aristotle on the Oblique," *Philosophical Studies* (Ire), 29: 89–101.

Englebretsen, G., 1985a, "Defending Distribution," *Dialogos*, 45: 157–159.

Englebretsen, G., 1985b, "Geach on Logical Syntax," *The New Scholasticism*, 59: 177–184,

Englebretsen, G., 1985c, "On the Proper Treatment of Negative Names," *The Journal of Critical Analysis*, 8: 109–115.

Englebretsen, G., 1985d, "Negative Names," *Philosophia*, 15: 133–136.

Englebretsen, G., 1985e, "Semantic Considerations for Sommers' Logic," *Philosophy Research Archives*, 11: 281–318.

Englebretsen, G., 1985f, "Quine on Aristotle on Identity," *Critica*, 17: 65–68.

Englebretsen, G, 1985 g, "Defending Distribution," *Dialogos*, 15: 157–159.

Englebretsen, G., 1986a, "Czezowski on Wild Quantity," *Notre Dame Journal of Formal Logic*, 27: 62–65.

Englebretsen, G., 1986b, "Singular/General," *Notre Dame Journal of Formal Logic*, 27: 104–107.
Englebretsen, G. 1987a, "Subjects," *Studia Leibnitiana*, 19: 85–90.
Englebretsen, G. (ed), 1987b, *The New Syllogistic*. New York and Bern: Peter Lang.
Englebretsen, G., 1988, "A Note on Leibniz's Wild Quantity Thesis," *Studia Leibnitiana*, 20: 87–89.
Englebretsen, G., 1990a, "The Myth of Modern Logic," *Cogito*, 4: 150–158.
Englebretsen, G., 1990b, "A Note on Copulae and Qualifiers," *Linguistic Analysis*, 20: 82–86.
Englebretsen, G., 1992a, "Linear Diagrams for Syllogisms (with Relationals)," *Notre Dame Journal of Formal Logic*, 33: 37–69.
Englebretsen G. 1992b, "An Algebra for Logic," *Canadian Journal of Rhetorical Studies*, 2: 104–140.
Englebretsen, G., 1992c, "Plus and Minus," *Critica*, 24: 73–116.
Englebretsen, G., 1996, *Something to Reckon With: The Logic of Terms*, Ottawa: University of Ottawa Press.
Englebretsen, G., 1997, "The Unifying Copula," *Logique et Analyse*, 159: 255–259.
Englebretsen, G., 1998, "Review: *The Logical Status of Diagrams* by Sun-Joo Shin," *Modern Logic*, 6: 322–330.
Englebretsen, F. 2000, "Preliminaries for a Term-Functor Logic," in: *Variable-Free Semantics*, M. Böttner and W. Thümmel (eds), Osnabrück: Secolo Verlag.
Englebretsen, G., 2002, "Syllogistic: Old Wine in New Bottles," *History and Philosophy of Logic*, 23: 31–35.
Englebretsen, G., 2004, "Predicate Logic, Predicates, and Terms," in: *First-Order Logic Revisited*, V. Hendricks, et al (eds), Berlin: Logos Verlag.
Englebretsen, G., 2005, "Trees, Terms, and Truth: The Philosophy of Fred Sommers," in: *The Old New Logic*, D. Oderberg (ed), Cambridge, MA: MIT Press.
Englebretsen, G., 2006, *Bare Facts and Naked Truths: A New Correspondence Theory of Truth*. Aldershot: Ashgate.
Englebretsen, G., 2010, "How to Use a Valid Derivers License," *The Reasoner*, 4: 54–55.
Englebretsen, G, 2012, *Robust Reality: An Essay in Formal Ontology*, Frankfurt: Ontos.
Englebretsen, G., 2015, *Exploring Topics in the History and Philosophy of Logic*, Berlin and Boston: de Gruyter.
Englebretsen, G., 2016a, "La Quadrature du Carré," in: *Soyons Logiques*, F. Schang, A. Moktefi, et A. Moretti (eds), London: College Publications.
Englebretsen, G., 2016b, "Fred Sommers' Contributions to Formal Logic," *History and Philosophy of Logic*, 37: 269–291.
Englebretsen, G. and C. Sayward, 2011, *Philosophical Logic: An Introduction to Advanced Topics*, London: Continuum.
Euler, L., 1768–1772, *Lettres à une Princesse d'Allemagne sur divers sujets de Physique et Philosophie*, Vols. 1 and 2, Saint Petersburg: Imprimerie de l'Académie Impériale des Sciences.
Euler, L., 1833, *Letters of Euler on Different Subjects in Natural Philosophy Addressed to a German Princess*, Vol. 1, New York: J. & J. Harper.
Franklin, J., 2000, "Diagrammatic Reasoning and Modelling in the Imagination: The Secret Weapons of the Scientific Revolution," in: *1543 and All That: Image and Word, Change*

and Continuity in the Proto-Scientific Revolution, G. Freeland and A. Corones (eds), Dordrecht, Kluwer, pp. 53–115.
Franklin, J., 2017, "Discrete and Continuous: A Fundamental Dichotomy in Mathematics," *Journal of Humanistic Mathematics*, 7: 355–378.
Frederick, D., 2013, "Singular Terms, Predicates and the Spurious 'is' of Identity," *Dialectica*, 67: 325–343.
Frederick, D., 2017, "The Unsatisfactoriness of Unsaturatedness," in: *Philosophy and Logic of Predication*, P. Stalmaszczyk (ed), New York and Bern: Peter Lang,
Frege, G., 1879, *Begriffsschrift: eine der arithmetischen nachgebildete Formelsprache des reinenDenkens*, Halle am See: Louis Nebert.
Frege, G., 1903, *Grundgesetze der Arithmetik*, vol 2, Jena: Pohl.
Frege, G., 1950, *Foundations of Arithmetic*, J.L. Austin (transl), Oxford: Basil Blackwell.
Frege, G, 1967, *The Basic Laws of Arithmetic*, M. Furth (transl and ed), Berkeley and Los Angeles: University of California Press
Frege, G., 1970, *Frege Translations*, P. Geach and M. Black (eds), Oxford: Basil Blackwell.
Frege, G., 1972, *Conceptual Notation and Related Articles*, T.W. Bynum (ed), Oxford: Oxford University Press.
Frege, G., 1979, *Posthumous Writings*, H. Hermes, F. Kambartel, and F. Kaulbach (eds), Oxford: Basil Blackwell.
Frege, G., 1980, *Philosophical and Mathematical Correspondence*, H. Hermes, F. Kambatel, C. Thiel and A. Veraart (eds), Chicago: University of Chicago Press.
Frege, G., 1984, *Collected Papers on Mathematics, Logic, and Philosophy*, B. McGuinness (ed), Oxford: Blackwell.
Frege, G. 1997, *The Frege Reader*, M. Beaney (ed), Oxford: Blackwell.
Friedman, W.H., 1978, "Uncertainties Over Distribution Dispelled," *Notre Dame Journal of Formal Logic*, 19: 653–662.
Gaines, B.R., 2010, "Visualizing Logical Aspects of Conceptual Structures," online: http://citeseerx.ist.psu.edu/viewdoc/download?doi=10.1.1.414.4316&red=red1&type=typepdf. Accessed December 5, 2017.
Gardner, M., 1982 [1958], *Logic Machines and Diagrams*, (2nd edition), Chicago: University of Chicago.
Geach, P., 1962, *Reference and Generality*, Ithaca, NY: Cornell University Press.
Geach, P., 1972, *Logic Matters*, Oxford: Blackwell.
Giaquinto, M., 2007, *Visual Thinking in Mathematics*, Oxford: Clarendon Press.
Giardino, V., 2013, "A Practice-Based Approach to Diagrams," in Shin and Moktefi, 2013.
Giardino, V. and Greenberg, G., 2015, "Introduction: Varieties of Iconicity," *Review of Philosophy and Psychology*, 6: 1–25.
Glasgow, J., Narayanan, N.H., and Chandrasekaran, B. (eds), 1995, *Diagrammatic Reasoning: Cognitive and Computational Perspectives*, Menlo Park, CA: AAAI Press and MIT Press.
Gottfried,B., 2014, "Set Space Diagrams," *Journal of Visual Language and Computing*, 25: 518–532.
Gottfried, B., 2015a, "A Comparative Study of Linear and Region Based Diagrams," *Journal of Information Science*, 10: 3–20.
Gottfried, B., 2015b, "The Diamond of Contraries," *Journal of Visual Language and Computing*, 26: 29–41.
Greaves, M., 2002, *The Philosophical Status of Diagrams*, Stanford: CSLI Publications.

Green, J., 2009, "The Problem of Elimination n the Algebra of Logic," in: *Perspective on the History of Mathematical Logic*, T. Drucker (ed), Basil and Boston: Birkhäuser.

Gurr, C., 1999, "Effective Diagrammatic Communications: Syntactic, Semantic and Pragmatic Issues," *Journal of Visual Languages and Computing*, 10: 317–342.

Gurr, C, Lee, J., and Stenning, K., 1998, "Theories of Diagrammatic Reasoning: Distinguishing Component Problems," *Minds and Machines*, 8: 533–557.

Hacking, I., 2007, "Trees of Logic, Trees of Porphyry," in: *Advancements of Learning: Essays in Honour of Paolo Rossi*, J.L. Heilbron (ed), Firenzie: L.S. Olschki, pp. 221–163.

Hamami, Y. and Mumma, J., 2012, "Euclid's Diagrammatic Logic and Cognitive Science," *CEUR Workshop Proceedings*, 883: 14–26.

Hammer, E., 1994, "Reasoning with Sentences and Diagrams," *Notre Dame Journal of Formal Logic*, 35: 73–87.

Hammer, E, 1995, *Logic and Visual Information*, Stanford: CSLI and FoLLI Publications.

Hammer, E., 1998, "Semantics for Existential Graphs," *Journal of Philosophical Logic*, 27: 489–503.

Hammer, E. and Shin, S.-J., 1998, "Euler's Visual Logic," *History and Philosophy of Logic*, 19: 1–29.

Hamilton, W.R., 1871, *Lectures on Logic*, Boston: Gould and Lincoln.

Hersh, R., 1997, *What is Mathematics, Really?* Oxford: Oxford University Press.

Hobbes, T., 1981, *Thomas Hobbes: Computatio Sive Logica, Part I of De Corpore*, I.C. Humgerland and G.R. Vick (eds), NY: Abris Books.

Hodges, W., 1998, "The Laws of Distribution for Syllogisms," *Notre Dame Journal of Formal Logic*, 39: 221–230.

Hodges, W., 2009, "Traditional Logic, Modern Logic and Natural Language," *Journal of Philosophical Logic*, 38: 589–606.

Hodges, W., 2018, "Two Early Arabic Applications of Model-Theoretic Consequence," *Logica Universalis*, online: https://doi.org/10.1007/s11787-018-0187–6. Accessed December 20, 2017.

Hoffmann, M.H.G., 2011, "Cognitive Conditions of Diagrammatic Reasoning," *Semiotica*, 189–212.

Horn, L.R., 1989, *A Natural History of Negation*, Chicago: University of Chicago Press; expanded and revised edition, 2001, Stanford Center for the Study of Language and Information.

Houser, N., Roberts, D.D., and J. Van Evra (eds), 1997, *Studies in the Logic of Charles Sanders Peirce*, Bloomington and Indianapolis: Indiana University Press.

Howell, R., 1976, "Ordinary Pictures, Mental Representations, and Logical Forms," *Synthese*, 33: 149–174.

Howse, J., Molina, F., Taylor, J., Kent, S., and Gil, J., 2001, "Spider Diagrams: A Diagrammatic Reasoning System," *Journal of Visual Languages and Computing*. 12: 299–324.

Howse, J., Stapleton, G., and Taylor, J., 2005, "Spider Diagrams," *London Mathematical Society Journal of Computation and Mathematics*, 8: 145–194.

Hubbeling, H.G., 1965, "A Diagram-Method in Propositional Logic," *Logique et Analyse*, 8: 227–288.

Hume, D., 1975, *A Treatise on Human Nature*, A. Selby-Bigge (ed), 2nd edition revised by P.H. Nidditch, Oxford: Oxford University Press.

Joaquin, J.J. and Boyles, R.J.M., "Teaching Syllogistic Logic via a Retooled Venn Diagram Technique," *Teaching Philosophy*.

John, C., Fish, A., Howse, J., and Taylor, J., 2006, "Exploring the Notion of Clutter in Euler Diagrams," 4th *International Conference on the Theory and Application of Diagrams*, Sprinker, pp. 267–282.

Johnson-Laird, P.N., 1983, *Mental Models: Towards a Cognitive Science of Language, Inference and Consciousness*, Cambridge, MA: Harvard University Press.

Johnson-Laird, P.N., 2002, "Peirce, Logic Diagrams, and the Elementary Operations of Reasoning," *Thinking and Reasoning*, 81: 69–95.

Johnson-Laird, P.N., 2010, "Mental Models and Human Reasoning," *Proceedings of the National Academy of Sciences*, 107: 18243–18250.

Johnson-Laird, P.N. and R.J. Byrn, 1991, *Essays in Cognitive Psychology: Deduction*, Hillsdale, NJ: Lawrence Erlbaum.

Kant, I., 1998, *Critique of Pure Reason*, P. Guyer and A. Wood (eds), Cambridge: Cambridge University Press.

Kanterian, E., 2012, *Frege: A Guide for the Perplexed*, London: Continuum.

Katz, B.D. and Martinich, A.P., 1976, "The Distribution of Terms," *Notre Dame Journal of Formal Logic*, 17: 279–283.

Ketner, K.L., (ed), 1995, *Peirce and Contemporary Thought: Philosophical Inquiries*, New York: Fordham University Press.

Khemlani, S. and Johnson-Laird, P.N., 2012, "Theories of the Syllogism: A Meta-Analysis," *Psychological Bulletin*, 138: 427–457.

Knauff, M., 2007, "How Our Brains Reason Logically," *Topoi*, 26: 19–36.

Kneale, W. and Kneale, M., 1962, *The Development of Logic*, Oxford: Clarendon Press.

Kuhn, S.T., 1983, "An Axiomatization of Predicate Functor Logic," *Notre Dame Journal of Formal Logic*, 24: 233–241.

Lambert, J.H., 1764, *Neuses Organon*, Leipzig: Wendler.

Lambert, J.H., 1771, *Anlage zur Architectonic, oder Theorie des Einfachen und des Ersten in der philosophischen und mathematischen Erkenntiniss*, vol. 1, Riga: J.F. Hartknoch.

Larkin, J. and Simon, H.A., 1995, "Why a Diagram is (Sometimes) Worth Ten Thousand Words," in: *Diagrammatic Reasoning: Cognitive and Computational Perspectives*, Glasgow, Narayanan, and Chandrasekaran (eds), Cambridge, MA: AAAI Press/MIT Press, 69–109. Original in *Cognitive Science*, 1987, 11: 65–99.

Laronge, J.A., 2012, "A Generalizable Argument Structure Using Defeasible Class-inclusion Transitivity for Evaluating Evidentiary Probative Relevancy in Litigation," *Journal of Logic and Computation*, 22: 129–162.

Legg, C., 2013, "What is a Logical Diagram?" in: *Visual Reasoning with Diagrams*, S.-J. Shin and A. Moktefi (eds), Basel: Springer, pp. 1–18.

Lehan-Streisel, V., 2012, "Why Philosophy Needs Logical Psychologism," *Dialogue*, 51: 575–586.

Leibniz, G.W., 1903, *Opuscules et fragments inédits de Leibniz: d'apres des manuscrits des docuements inédits*, L. Couturat (ed), Paris: Félix Alcan.

Leibniz, G.W., 1949, *New Essays Concerning Human Understanding*, La Salle, IL: Open Court Publishing.

Leibniz, G.W., 1966, *Logical Papers*, G.H.R. Parkinson (transl and ed), Oxford: Clarendon Press.

Lemanski, J., 2016, "Means or End? On the Valuation of Logic Diagrams," *Logico-Philosophical Studies*, 14: 98–121.
Lemanski, J., 2017, "Periods in the History of Euler-Type Diagrams," *Acta Baltica Historiae et Philosophiae Scientiarum*, 5: 50–69.
Lemanski, J., 2018, "Logic Diagrams in the Weigel and Weise Circles," *History and Philosophy of Logic*, 39: 3–28.
Lemon, A. and Lemon, O., 2000, "Constraint Matching for Diagram Design: Qualitative Visual Languages," in: *Diagrams 2000*, M. Anderson, P. Cheng, and V. Haarslev (eds), Berlin: Springer, pp. 74–88.
Lemon, D., 2017, *Drawing Physics*, Cambridge, MA: MIT Press.
Lemon, O., 1997, "Review: *Logic and Visual Information*, Eric M. Hammer," *Journal of Logic, Language, and Information*, 6: 213–216.
Lemon, O., 2002, "Comparing the Efficacy of Visual Languages," in: *Words, Proofs, and Diagrams*, Barker-Plummer, D., Beaver, D., van Benthem, J., and di Luzio, P.S. (eds), 2002, Stanford: CSLI Publications, pp. 47–69.
Lemon, O., De Rijke, M., and Shimojima, A., 1999, "Editorial: Efficacy of Diagrammatic Reasoning," *Journal of Logic, Language and Information*, 8: 265–271.
Lemon, O. and Pratt, I., 1997, "Spatial Logic and Complexity of Diagrammatic Reasoning," *Machine Graphics and Vision*, 6: 89–108.
Lemon, O. and Pratt, I., 1998, "On the Insufficiency of Linear Diagrams for Syllogisms," *Notre Dame Journal of Formal Logic*, 39: 573–580.
Lemon, O. and Pratt, I., 1999, "Drawing Illusions – A Case Study in the Incorrectness of Diagrammatic Reasoning," University of Dublin, Technical Report, online: www.tara.ie/bitstream/handle/2262/13012/TCD-CS-1999-17.pdf?sequence=1. Accessed November 25, 2017.
Lenzen, W., 1989, "Concepts vs Predicates: Leibniz's Challenge to Modern Logic," Firenze: Cenro Florentio di Storia e Filosofia della Scienza, pp. 153–172.
Lindsay, R.K., 1995, "Imagery and Inference," in: *Diagrammatic Reasoning: Cognitive and Computational Perspectives*, Chandrasekaran, B., Glasgow, J., and Narayanan, N. (eds), Cambridge, MA: AAAI Press/MIT Press.
Linnebo, Ø., Shapiro, S., and Hellman, G., 2016, "An Aristotelian Continuum," *Philosophia Mathematica*, 24: 214–246.
Lockwood, M., 1975, "On Predicating Proper Names," *Philosophical Review*, 84: 471–498.
Łukasiewicz, J., 1957, *Aristotle's Syllogistic from the Standpoint of Modern Formal Logic*, 2nd edition, Oxford: Oxford University Press.
Macbeth, D., 2005, *Frege's Logic*, Cambridge, MA: Harvard University Press.
Macbeth, D., 2012a, "Diagrammatic Reasoning in Frege's *Begriffsschrift*," *Synthese*, 186: 289–314.
Macbeth, D., 2012b, "Seeing How it Goes: Paper-and-Pencil Reasoning in Mathematical Practice," *Philosophia Mathematica*, 20: 58–85.
Macbeth, D., 2013, "Writing Reason," *Logique et Analyse*, 221: 25–44.
MacColl, H., 1880, "Symbolical Reasoning," *Mind*, 5: 45–60.
Macnamara, J. 1986, *A Border Dispute: The Place of Logic in Psychology*, Cambridge, MA: MIT Press.
MacQueen, G., 1967, The Logic Diagram, MA thesis, McMasters University.

Magnani, L., 2015, "Naturalizing Logic: Errors of Reasoning Vindicated: Logic Reapproaches Cognitive Science," *Journal of Applied Logic*, 13: 13–36.

Makinson, D., 1969, "Remarks on the Concept of Distribution in Traditional Logic," *Noûs*, 3: 103–108.

Marquand, A., 1881, "On Logical Diagrams for N Terms," *Philosophical Magazine*, series 5, 12: 266–290.

Martin, J.N., 2013, "Distributive Terms and the Port Royal Logic," *History and Philosophy of Logic*, 34: 133–154.

Masthoff, J. and G. Stapleton, 2007, "Incorporating Negation into Visual Logics: A Case Study Using Euler Diagrams," online: http://homepages.abdn.ac.uk/j.masthoff/pages/Publications/VLC07.pdf. Accessed December 5, 2017.

Merrill, D.D., 1997, "Relations and Quantification in Peirce's Logic," in: *Studies in the Logic of Charles Sanders Peirce*, N. Houser, N., D.D. Roberts, and J. Van Evra (eds), 1997, Peirce, Bloomington and Indianapolis: Indiana University Press.

Mineshima, K., Okada, M., and Takemura, R., 2009, "Conservativity for a Hierarchy of Eulerand Venn Reasoning Systems," *Proceedings of Visual Languages and Logic 2009, CEUR Series*, 510: 37–61.

Mineshima, K., Okada, M., and Takemura, R., 2010, "Two Types of Diagrammatic Inference Systems: Natural Deduction Style and Resolution Style," in: *Diagrams 2010, LNIA 6179*, A.K. Goel, M. Jamnick, and N.H. Narayanan (eds), pp. 99–114, online: http://abelard.flet.keio.ac.jp/person/takemura/index.html. Accessed December 10, 2017.

Mineshima, K., Okada, M., and Takemura, R., 2012, "A Generalized Syllogistic Inference System Based on Inclusion and Exclusion," *Studia Logica*, 100: 753–785.

Mineshima, K., Sato, Y., Takemura, R., and Okada, M., 2014, "Toward Explaining the Cognitive Efficacy of Euler Diagrams in Syllogistic Reasoning: A Relational Perspective," *Journal of Visual Languages and Computing*, 25: 156–165.

Moktefi, A., 2013, "Beyond Syllogisms: Carroll's (Marked) Quadrilateral Diagram," in: *Visual Reasoning with Diagrams*, S.-J. Shin and A. Moktefi (eds), Basel: Springer, pp. 55–71.

Moktefi, A., 2015, "Is Euler's Circle a Symbol or an Icon?" *Sign System Studies*, 43: 697–615.

Moktefi, A., 2017, "Diagrams as Scientific Instruments," in: *Visual, Virtual, Veridical*, A. Benedek and A. Veszelszki (eds), Frankfurt: Peter Lang, pp. 81–89.

Moktefi, A. and Edwards, A.W.F., 2011, "One More Class: Martin Gardner and Logic Diagrams," in: *A Bouquet for Gardner: Martin Gardner Remembered*, M. Burstein (ed), New York: Lewis Carroll Society of North America.

Moktefi, A., and Pietarinen, A.-V., 2016a, "On the Diagrammatic Representation of Existential Statements with Venn Diagrams," *Journal of Logic, Language, and Information*, 24: 361–374.

Moktefi, A., and Pietarinen, A.-V., 2016b, "Negative Terms in Euler Diagrams: Peirce's Solution," in: *Diagrammatic Representation and Inference: Diagrams 2016*, M. Jamnik et al (eds), Basil: Springer.

Moktefi, A. and Shin, S.-J., 2012, "A History of Logic Diagrams," in: *Logic: A History of its Central Concepts*, D.M. Gabby, F.J. Pelletier, and J. Woods (eds), Amsterdam: North-Holland, pp. 611–682.

Moktefi, A., Bellucci, F., and Pietarinen, A.-V., 2014, "Continuity, Connectivity and Regularity in Spatial Diagrams for N Terms," in: *Proceeding of the First International Workshop on*

Diagrams, Logic and Cognition, J. Burton and L. Choudhury (eds), *CEUR Workshop Proceedings*, 1132: 31–35, online: http://ceur-ws.org/Vol-1132/. Accessed November 20, 2017.

Morrissey, C.S., 2015, "A Logic Without Nominalism: Existential Assumptions on the Aristotelian Square of Opposition Revisited," *American Journal of Semiotics*, 31: 183–202.

Moss, L.S., 2015, "Natural Logic," in: *The Handbook of Contemporary Semantic Theory*, S. Lappin and C. Fox (eds), Chichester, UK: Wiley.

Mostowski, A., 1957, "On a Generalization of Quantifiers, *Fundamenta Mathematicae*, 44: 12–36.

Mozes, E., 1989, "A Deductive Database Based on Aristotelian Logic," *Journal of Symbolic Computation*, 7: 487–507.

Murphree, W.A., 1991, *Numerically Exceptive Logic: A Reduction of the Classical Syllogism*, NY: Peter Lang.

Murphree, W.A., 1997, "The Numerical Syllogisms and Existential Presupposition," *Notre Dame Journal of Formal Logic*, 38: 49–64.

Murphree, W.A., 1998, "Numerical Term Logic," *Notre Dame Journal of Formal Logic*, 39: 346–362.

Natali, C., 2013, *Aristotle: His Life and School*, Princeton: Princeton University Press.

Nakatsu, R.T., 2009, *Diagrammatic Reasoning in AI*, Wiley, online: DOI: 10.1002/9780470400777. Accessed December 4, 2017.

Nersessian, N.J., 1992, "In the Theoretician's Laboratory: Thought Experimenting as Mental Modeling," *PSA: Proceeding of the Biennial Meeting of the Philosophy of Science Association*, 2: 291–301.

Noah, A., 1980, "Predicate-Functors and the Limits of Decidability in Logic," *Notre Dame Journal of Formal Logic.* 21: 701–707.

Noah, A., 1982, "Quine's Version of Term Logic," Appendix E of *The Logic of Natural Language*, F. Sommers, 1982, Oxford: Clarendon Press.

Noah, A., 1987, "The Two Term Theory of Predication," in: *The New Syllogistic*, G. Englebretsen (ed), 1987, New York: Peter Lang.

Noah, A., 1993, "Nonclassical Syllogistic Inference and the Method of Resolution," *Notre Dame Journal of Formal Logic*, 34: 209–222.

Noah, A., 2005, "Sommers' Cancellation Technique and the Method of Resolution," in: *The Old New Logic: Essays on the Philosophy of Fred Sommers*, D. Oderberg (ed), 2005, Cambridge, MA: MIT Press.

Noah, G., 1973, *Singular Terms and Predication*, Brandeis University, PhD thesis.

Norman, D., 1993, "Cognition in the Head and in the World," *Cognitive Science*, 17: 1–6.

Nùñez, R., 2017, "Is There Really an Evolved Capacity for Number?" *Trends Cognitive Science*, 21: 409–424.

Osherson, D.N., 1995, "Probability Judgment," in: *Thinking*, E.E. Smith and D.N. Osherson (eds), Cambridge, MA: MIT Press, pp. 35–75.

Pagnan, R., 2010, "A Diagrammatic Calculus of n-Term Syllogisms," online: arXiv:1001.1707v3 [math.LO]. Accessed December 16, 2017.

Pagnan, R., 2012, "A Diagrammatic Calculus of Syllogisms," *Journal of Logic, Language and Information*, 21: 347–364.

Pagnan, R., 2013a, "A Diagrammatic Calculus of Syllogisms," in: *Visual Reasoning with Diagrams*, S.-J. Shin and A. Moktefi (eds), Basel: Springer, pp. 33–53.

Pagnan, R., 2013b, "Syllogisms in Rudimentary Linear Logic, Diagrammatically," *Journal of Logic, Language and Information*, 22: 71–113, online: http://link.springer.com/article/10.1007%Fs10849-123-9170-4#page-1. Accessed December 9, 2017.

Parkinson, G.H.R., 1966, "Introduction," *Leibniz: Logical Papers* G.H.R. Parkinson (transl and ed), Oxford: Clarendon Press, pp. ix-lxv.

Parsons, T., 2006, "The Doctrine of Distribution," *History and Philosophy of Logic*, 27: 59–74.

Parsons, T., 2015, *Articulating Medieval Logic*, Oxford: Oxford University Press.

Patzig, G., 1968, *Aristotle's Theory of the Syllogism*, Dordrecht: Reidel.

Peirce, C.S., 1870, "Description of a Notation for the Logic of Relatives, Resulting from an Amplification of the Conceptions of Booles's Calculus of Logic," *Memoirs of the Academy of Arts and Sciences*, 9: 317–378.

Peirce, C.S., 1885, "On the Algebra of Logic: A Contribution to the Philosophy of Notation," *American Journal of Mathematics*, 7: 180–202.

Peirce, C.S., 1897, "The Logic of Relatives," *The Monist*, 7: 161–217.

Peirce, C.S., 1892, "The Critic of Arguments, II (The Reader is Introduced to Relatives)," *The Open Court*, 6: 3416–3418; reprinted in: Peirce 1931–1958, vol. 4.

Peirce, C.S., 1931–1958, *Collected Papers of Charles Sanders Peirce*, C. Hartshorne, P. Weiss and A.W. Burks (eds), Cambridge, MA: Harvard University Press.

Peirce, C.S. 1998, *The Essential Peirce*, N. Houser, et al (eds), Bloomington: Indiana University Press.

Pelletier, F.J., 2001, "Did Frege Believe Frege's Principle?" *Journal of Logic, Language, and Information*, 10: 87–114.

Pelletier, F.J., Elio, R., and Hanson, P., 2008, "Is Logic All in Our Heads?" *Studia Logica*, 86: 1–65.

Pereira-Fariña, M., 2014, "Some Reflections on the Set-based and the Conditional-based Interpretation of Statements in Syllogistic Reasoning," *Archives for the Philosophy and History of Soft Computing*, 1: 1–16.

Peterson, P.L., 1979, "On the Logic of 'Few', 'Many', and 'Most'," *Notre Dame Journal of Formal Logic*, 20: 155–179.

Peterson, P.L., 1985, "Higher Quantity Syllogisms," *Notre Dame Journal of Formal Logic*, 26: 348–360.

Peterson, P.L. 1988, *Syllogistic Logic and the Grammar of Some English Quantifiers*. Indianapolis: Indiana University Linguistics Club.

Peterson, P.L., 1991, "Complexly Fractionated Syllogistic Quantifiers," *Journal of Philosophical Logic*, 20: 287–313.

Peterson, P.L., 1993, "Intermediate Quantifiers for Finch's Propositions," *Notre Dame Journal of Formal Logic*, 34: 140–149.

Peterson, P.L., 1996, "Review: *The Logical Status of Diagrams* by Sun-Joo Shin," *Canadian Philosophical Reviews*, 16: 208–210.

Peterson, P.L., 2000, *Intermediate Quantifiers: Logic, Linguistics, and Aristotelian Semantics*, Aldershot: Ashgate.

Pietarinen, A.-V., 2011, "Existential Graphs: What a Diagrammatic Logic of Cognition Might Look Like, *History and Philosophy of Logic*, 32: 265–281.

Pietarinen, A.-V. and Bellucci, F., 2016, "The Iconic Moment: Towards a Peircean Theory of Diagrammatic Imagination," in: *Epistemology, Knowledge and the Impact of Interaction*, J. Redman et al (eds), Basil: Springer.
Pinheiro, F.F., 2016, "Correção Gráfica e Outras Características do Método Diagramátco 'Díagrfos de Gardner'," ("Graphic Soundness and Other Characteristics of the Diagrammatic Method 'Gardner's Diagraphs'" *Kínesis – Revista de Estudos dos Pós-Graduandos em Filosofia*, 7: 245–258.
Plato, 1937, "The Republic," *The Dialogues of Plato*, B. Jowett (transl), Oxford: Oxford University Press.
Plato, 1968, *The Republic of Plato*, Bloom, A. (transl), 2nd edition, New York: Basic Books.
Powell, A., 2003, "Visualisation and Logic," *Teorema*, 22: 95–104.
Putnam, H., 1982, "Peirce the Logician," *Historia Mathematica*, 9: 290–301.
Putnam, H., 1995, "Peirce's Continuum," in: *Peirce and Contemporary Thought: Philosophical Inquiries*, K.L. Ketner (ed), New York: Fordham University Press.
Quine, W.V.O., 1936a, "Concepts of Negative Degree," *Proceeding of the National Association of Science*, 22: 40–45.
Quine, W.V.O., 1936b, "Towards a Calculus of Concepts," *Journal of Symbolic Logic*, 1: 2–25.
Quine, W.V.O., 1937, "Logic Based on Inclusion and Abstraction," *Journal of Symbolic Logic*, 2: 145–152.
Quine, W.V.O., 1959, "Eliminating Variables Without Applying Functions to Functions, *Journal of Symbolic Logic*, 24: 324–325.
Quine, W.V.O., 1960a, "Variables Explained Away," *Proceedings of the American Philosophical Society*, 104: 343–347.
Quine, W.V.O., 1960b, *Word and Object*, Cambridge, MA: MIT Press.
Quine, W.V.O., 1971, "Predicate Functor Logic," in: *Proceedings of the Second Scandinavian Logic Symposium*, J. Fenstand (ed), Amsterdam: North-Holland Publ.
Quine, W.V.O., 1976a, "The Variable," *The Ways of Paradox and Other Essays*, revised edition, W.VO. Quine, Cambridge, MA: Harvard University Press.
Quine, W.V.O., 1976b, "Algebraic Logic and Predicate Functors," *The Ways of Paradox and Other Essays*, revised edition, W.VO. Quine, Cambridge, MA: Harvard University Press.
Quine, W.V.O., 1981a, "Predicate Functors Revisited," *Journal of Symbolic Logic*, 46: 649–652.
Quine, W.V.O., 1981b, "Predicates, Terms, and Classes," *Theories and Things*, W.V.O. Quine, Cambridge, MA: Harvard University Press.
Ramsey, F. 1925, "Universals," *Mind*. 34: 401–417.
Rauf, J.V., 1996, *Math Matters*, NY: Wiley.
Read, S., 2015, "Aristotle and Łukasiewicz on Existential Import," *Journal of the American Philosophical Association*, 1: 535–544.
Rearden, M., 1984, "The Distribution of Terms," *The Modern Schooman*, 61: 187–195.
Rips, L., 1994, *The Psychology of Proof: Deductive Reasoning in Human Thinking*, Cambridge, MA: MIT Press.
Rips, L., 1995, "Deduction and Cognition," in: *Thinking: An Invitation to Cognitive Science*, Smith, E.E. and Osherson, D.N., (eds), 1995, Cambridge, MA: MIT Press, pp. 297–343.
Roberts, D.D., 1973, *The Existential Graphs of Charles S. Peirce*, The Hague: Mouton.
Roeper, P., 2006, "The Aristotelian Continuum: A Formal Characterization," *Notre Dame Journal of Formal Logic*, 47: 211–231.

Rogers, P., Stapleton, G., and Chapman, P., 1015, "Visualizing Sets with Linear Diagrams," *ACM Transactions on Computer-Human Interaction*, 22: 1–39.

van Rooij, R., 2012, "The Propositional and Relational Syllogistic," *Logique et Analyse*, 55: 85–101.

Ross, W.D., 1949, "Commentary," in: *Aristotle's Prior and Posterior Analytics*. W.D. Ross (ed), Oxford: Oxford University Press.

Ross, W.D., 1960, *Aristotle: A Complete Exposition of his Work and Thought*, 2nd edition, New York: Meridian.

Russell, B., 1903, *The Principles of Mathematics*. London: Allen & Unwin.

Russell, B., 1918, *The Philosophy of Logical Atomism*, reprinted in: *Logic and Knowledge*, R.C. Marsh (ed), London: Allen & Unwin, 1956.

Rybak, J. and Rybak, J., 1976, "Venn Diagrams Extended: Map Logic," *Notre Dame Journal of Formal Logic* 17: 469–475.

Rybak, J. and Rybak, J., 1984a, "Mechanizing Logic I: Map Logic Extended Formally to Relational Arguments," *Notre Dame Journal of Formal Logic*, 25: 250–264.

Rybak, J. and Rybak, J., 1984b, "Mechanizing Logic II: Automated Map Logic Method for Relational Arguments on Paper and by Computer," *Notre Dame Journal of Formal Logic*, 25: 265–282.

Ryle, G., 1949, *The Concept of Mind*, London: Hutchinson University Library.

Sato, Y., 2013, *The Cognitive Efficacy of Diagrammatic Representations in Logical Reasoning*, PhD thesis, Keio University.

Sato, Y. and Mineshima, K., 2012, "The Efficacy of Diagrams in Syllogistic Reasoning: A Case for Linear Diagrams (Diagrams), *Lecture Notes in Computer Science*, 7352: pp. 352–355.

Sato, Y. and Mineshima, K., 2015, "How Diagrams Support Syllogistic Reasoning: An Experimental Study," *Journal of Logic, Language and Information*, 24: 409–455.

Sato, Y. and Mineshima, K., 2016, "Human Reasoning with Proportional Quantifiers and its Support by Diagrams," *Diagrammatic Representation and Inference, Lecture Notes in Computer Science*, 9781: 123–138.

Sato, Y., Mineshima, K. and Takemura, R., 2010a, "The Efficacy of Euler and Venn Diagrams in Deductive Reasoning: Empirical Findings," *Lecture Notes in Computer Science*, 6170: 6–22.

Sato, Y., Mineshima, K. and Takemura, R., 2010b, "Constructing Internal Diagrammatic Proofs from External Logic Diagrams" in: *Proceedings of the 32nd Annual Conference of the Cognitive Science Society*, R. Catrambone and S. Ohlsson (eds), Austin, TS: Cognitive Science Society.

Sato, Y., Mineshima, K. and Takemura, R., 2011, "Interpreting Logic Diagrams: A Comparison of Two Formulations of Diagrammatic Representations," *Proceedings of the 33rd Annual Conference of the Cognitive Science Society*, pp. 2182–2187.

Sato, Y., Wajima, Y., and Ueda, K., 2014, "Visual Bias of Diagram in Logical Reasoning," *Proceedings of the Annual Meeting of the Cognitive Science Society*, 36: 2883–2888.

Savio, M., 1998, "AE (Aristotle-Euler) Diagrams: An Alternative Complete Method for the Categorical Syllogism," *Notre Dame Journal of Formal Logic*, 39: 581–599.

Scaife, M. and Rogers, Y., 1996, "External Cogitation: How do Graphical Representations Work?" *International Journal of Human-Computer Studies*, 45: 185–213.

Schang, F. and Moktefi, A., 2008, "Depicting Negation in Diagrammatic Logic: Legacy and Prospects," *International Conference on Theory and Application of Diagrams*, Berlin and Heidelberg: Springer.

Schlimm, D., 2017, "Frege's Begriffsschrift Notation for Propositional Logic," *History and Philosophy of Logic*, 39:53–79, online: http://dx.doi.org/10.1080/01445340,2017.1317429. Accessed November 24, 2017.

Schröder, E., 1880. "Gottlob Frege, *Begriffsschrift*," *Zeitschrift für Mathematik und Physik*, 25: 82–94.

Segerberg, K., 2000, *A Carrollian Introduction to Categorical Logic*, Uppsala: Uppsala University.

Shapiro, S., and Hellman, G., 2015, "Frege Meets Aristotle: Points as Abstracts," *Philosophia Mathematica*, online: http://philmat.oxfordjournals.org. Accessed November 5, 2017.

Sharp, D., Côté, M.-H., and Eakin, L., 1999, "Reasoning about a Structured Object: Three- and Four-Year-Olds' Grasp of a *Borderline Case* and an *Unexcluded Middle*." *Child Development*, 70: 866–871.

Shimojima, A., 1996a, "Operational Constraints in Diagrammatic Reasoning," in: *Logical Reasoning with Diagrams*, G. Allwein and J. Barwise (eds), NY: Oxford University Press, 1996, pp. 27–48.

Shimojima, A., 1996b, "Reasoning with Diagrams and Geometric Constraints," in: *Logic, Language and Conceptual Computation*. Vol. 1, J. Seligman and D. Westerstahl (eds), Stanford: CSLI Publications, pp. 527–540.

Shimojima, A., 2015, *Semantic Properties of Diagrams and Their Cognitive Potentials*, Stanford: CSLI Publications.

Shin, S.-J., 1994a, *The Logical Status of Diagrams*, New York: Cambridge University Press.

Shin, S.-J., 1994b, "Peirce and the Logical Status of Diagrams," *History and Philosophy of Logic*, 15: 45–68.

Shin, S.-J., 2002, *The Iconic Logic of Peirce's Graphs*, Cambridge, MA: MIT Press.

Shin, S.-J., 2004, "Heterogeneous Reasoning and its Logic," *Bulletin of Symbolic Logic*, 10: 86–106.

Shin, S.-J. and Hammer, E., 2014, "Peirce's Deductive Logic," in: *Stanford Encyclopedia of Philosophy*, E. Zalta (ed), online: httpi//plato.stanford.edu/archives/fall2014/entries/peirce-logic/. Accessed December 8, 2017.

Shin, S.J., Lemon, O., and Mumma, J., 2016, "Diagrams," in: *The Stanford Encyclopedia of Philosophy*, E.N. Zalta (ed), online: http://platostanford.edu/archives/win2016/entries/diagrams/. Accessed December 17, 2017.

Shin, S.-J. and Moktefi, A., (eds), 2013, Visual Reasoning with Diagrams, Basel: Springer.

Skosnik, J., 1980, "Leibniz and Russell on Existence and Quantification Theory," *Canadian Journal of Philosophy*, 4: 681–720.

Slater, B.H., 1998, "Peirce's Graphs Amended," *History and Philosophy of Logic*, 19: 101–106.

Sluga, H., 1987, "Frege Against the Booleans," *Notre Dame Journal of Formal Logic*, 28: 80–98.

Smiley, T., 1973, "What is a Syllogism?" *Journal of Philosophical Logic*, 2: 136–154.

Smiley, T., 1994, "Aristotle's Completeness Proof," *Ancient Philosophy*, 14: 24–38.

Smit, H., 2018, "Inclusive Fitness Theory and the Evolution of Mind and Language," *Erkenntnis*, 83: 287–314.
Smit, H. and Hacker, P.M.S., 2014, "Seven Misconceptions About the Mereological Fallacy: A Compilation for the Perplexed," *Erkenntnis*, 79, 1077–1097.
Smith, E.E. and Osherson, D.N., (eds), 1995, *Thinking: An Invitation to Cognitive Science*, Cambridge, MA: MIT Press.
Smyth, M.B., 1971, "A Diagrammatic Treatment of Syllogistic," *Notre Dame Journal of Formal Logic*, 12: 483–488.
Sober, E., 1976, "Mental Representations," *Synthese*, 33: 101–148.
Sommers, F., 1967, "On a Fregean Dogma," in: *Problems in the Philosophy of Mathematics*, I. Lakatos (ed), Amsterdam: North-Holland Publ.
Sommers, F., 1969, "Do We Need Identity?" *Journal of Philosophy*, 66: 499–504.
Sommers, F., 1970, "The Calculus of Terms," *Mind*, 79: 1–39. Reprinted in: *The New Syllogistic*, G. Englebretsen (ed), 1987, New York: Peter Lang.
Sommers, F., 1973, "The Logical and the Extra-Logical," *Boston Studies in the Philosophy of Science*, 14: 235–252.
Sommers, F., 1975, "Distribution Matters," *Mind*, 84: 27–46.
Sommers, F., 1976a, "Leibniz's Program for the Development of Logic," in: *Essays in Memory of Imre Lakatos*, R.S. Cohen et al (eds), Dordrecht: D. Reidel.
Sommers, F., 1976b, "Frege or Leibniz," in: *Studies on Frege III*, M. Schirn (ed), Stuttgart: Fromman-Holzboog.
Sommers, F., 1976c, "Logical Syntax in Natural Language", in: *Issues in the Philosophy of Language*, A.F. MacKay and D.D. Merrill (eds), New Haven and London: Yale University Press.
Sommers, F., 1976d, "On Predication and Logical Syntax," in: *Language in Focus*, A. Kasher (ed), Dordrecht: D.Reidel.
Sommers, F., 1978, "The Grammar of Thought," *Journal of Social and Biological Structures*, 1: 39–51,
Sommers, F., 1981, "Are There Atomic Propositions?" *Midwest Studies in Philosophy*, 6: 59–68.
Sommers, F., 1982, *The Logic of Natural Language*, Oxford: Clarendon Press.
Sommers, F., 1983a, "Linguistic Grammar and Logical Grammar," in: *How Many Questions? Essays in Honor of Sidney Morgenbesser*, L.S. Cauman et al (eds), Indianapolis: Hackett.
Sommers, F., 1983b, "Reply to Geach," *Times Literary Supplement*, 14 January.
Sommers, G., 1983c, "Reply to Geach," *Times Literary Supplement*, 18 February.
Sommers, F., 1983d, "The Grammar of Thought: A Reply to Dauer," *Journal of Social and Biological Structures*, 6: 37–44.
Sommers, F., 1987, "Truth and Existence," in: *The New Syllogistic*, G. Englebretsen (ed), New York: Peter Lang.
Sommers, F., 1990, "Predication in the Logic of Terms," *Notre Dame Journal of Formal Logic*, 31: 106–126.
Sommers, F., 1993, "The World, the Facts, and Primary Logic," *Notre Dame Journal of Formal Logic*, 34: 169–182.
Sommers, F., 2000, "Term Functor Grammars," in: *Variable-Free Semantics*, M. Böttner and W. Thümmel (eds), Osnabrück: Secolo Verlag.

Sommers, F., 2002, "On the Future of Logic Instruction," *American Philosophical Association Newsletter on Teaching Philosophy*, 01: 176–180.
Sommers, F., 2005a, "Bar-Hillel's Complaint," *Philosophia*, 33: 55–68.
Sommers, F. 2005b, "Comments and Replies," in: *The Old New Logic: The Philosophy of Fred Sommers*, D. Oderberg (ed), Cambridge, MA: MIT Press.
Sommers, F. 2005c, "Intellectual Autobiography," in: *The Old New Logic: The Philosophy of Fred Sommers*, D. Oderberg (ed), Cambridge, MA: MIT Press.
Sommers, F., 2008a, "Reasoning: How We're Doing It," in: *The Reasoner*, 2: 5–7.
Sommers, F., 2008b, "Ratiocination: An Empirical Account," *Ratio*, 21: 115–133 l. Reprinted in: G. Englebretsen, 2015, *Exploring Topics in the History and Philosophy of Logic*, Berlin and Boston: De Gruyter.
Sommers, F., 2008c, "Ryle's Way with the Liar," in: Englebretsen, 2015, *Exploring Topics in the History and Philosophy of Logic*, Berlin and Boston: De Gruyter.
Sommers, F. and Englebretsen, G., 2000, *An Invitation to Formal Reasoning*, Aldershot: Ashgate.
Sowa, J., 1984, *Conceptual Structures: Information Processing in Mind and Machine*, Boston: Addison-Wesley Longman.
Spelke, E., Lee, S.A., and Izard, V., 2010, "Beyond Core Knowledge: Natural Geometry," *Cognitive Science*, 34: 863–884.
Sperber, D., and Mercier, H., 2017, *The Enigma of Reason: A New Theory of Human Understanding*, Cambridge, MA: Harvard University Press.
Stapleton, G., Blake, A., Burton, J., and Touloumis, A., 2017, "Presence and Absence of Individuals in Diagrammatic Logics: An Empirical Comparison," *Studia Logica*, 105: 787–815.
Stapleton, G., Howse, J, Taylor, J., and Thompson, S., 2004a, "The Expressiveness of Spider Diagrams," *Journal of Logic Computation*, 14: 857–880.
Stapleton, G., Howse, J., Taylor, J., and Thompson, S., 2004b, "What Can Spider Diagrams Say?" in: *Diagrams 2004*, A. Blackwell, et al (eds), Berlin: Springer.
Stapleton, G., Howse, J., Thompson, S., Taylor, J., and Chapman, P., 2013, "On the Completeness of Spider Diagrams Augmented with Constants," in: *Visual Reasoning with Diagrams*, S.-J. Shin and A. Moktefi (eds), Basel: Springer, pp. 101–133.
Stapleton, G., Jamnik, M., and Shimojima, A., 2017, "What Makes an Effective Representation of Information: A Formal Account of Observational Advantages," *Journal of Logic, Language, and Information*, 26: 143–177.
Stapleton, G. and Masthoff, J., 2007, "Incorporating Negation into Visual Logics: A Case Study Using Euler Diagrams," *Visual Languages and Computing*, pp. 187–194.
Stenning, K., 2002, *Seeing Reason: Image and Language in Learning to Think*, Oxford: Oxford University Press.
Stenning, K., Cox, R., and Oberlander, J., 1995a, "Contrasting the Cognitive Effects of Graphical and Sentential Logic Teaching: Reasoning, Representation and Individual Differences," *Language and Cognitive Processes*, 10: 333–354.
Stenning, K., Cox, R., and Oberlander, J., 1995b, "The Effect of Graphical and Sentential Logic Teaching on Spontaneous External Representation," *Cognitive Studies: Bulletin of the Japanese Cognitive Science Society*, 2: 56–75.
Stenning, K. and Lemon, O., 2001, "Aligning Logical and Psychological Perspectives on Diagrammatic Reasoning," *Artificial Intelligence Review*, 15: 29–62.

Stenning, K. and Oberlander, J., 1995, "A Cognitive Theory of Graphical and Linguistic Reasoning: Logic and Implementation," *Cognitive Science*, 19: 97–140.

Stenning, K. and van Lambalgen, M., 2008, *Human Reasoning and Cognitive Science*, Cambridge, MA: MIT Press.

Stjernfelt, F., 2007, *Diagrammatology: An Investigation on the Borderlines of Phenomenology, Ontology, and Semiotics*, Dordrect: Springer.

Strawson, P.F., 1957, "Logical Subjects and Physical Objects," *Philosophy and Phenomenological Research*, 17: 441–457.

Strawson, P.F., 1970, "The Asymmetry of Subjects and Predicates," in: *Language, Belief, and Metaphysics*, H.E. Kiefer and M. Munitz (eds), Albany: State University of New York, pp. 69–86.

Strawson, P.F., 1974, *Subject and Predicate in Logic and Grammar*, London: Methuen.

Szabó, Á., 1978, *The Beginnings of Greek Mathematics*, Dordrecht and Boston: D. Reidel.

Szabolcsi, L., 2008, *Numerical Term Logic*, Lewiston, NY: Mellen.

Tennant, N., 1986, "The Withering Away of Formal Semantics?" *Mind & Language*, 1: 302–318.

Toader, I. D., 2004, "Frege's Logical Diagrams," in: *Diagrammatic Representation and Inference: International Conference on Theory and Application of Diagrams*, A. Blackwell, K. Marriot, and A. Shimojima (eds), Berlin Heidelberg: Springer.

Thom, P., 1977, "*Termini Obliui* and the Logic of Relations," *Archiv für Feschichte der Philosophie*, 59: 143–155.

Thompson, B., 1982, "Syllogisms Using 'few', 'many', and 'most'," *Notre Dame Journal of Formal Logic*, 23: 75–84.

Thompson, B., 1986, "Syllogisms with Statistical Quantifiers," *Notre Dame Journal of Formal Logic*, 27: 93–103.

Thompson, B., 1992, *An Introduction to the Syllogism: and the Logic of Proportional Quantifiers*, NY: Peter Lang.

Veatch, H.B., 1952, *Intentional Logic*, New Haven: Yale University Press.

Venn, J., 1880, "On the Diagrammatic and Mechanical Representation of Propositions and Reasonings," *The London, Edinburgh and Dublin Philosophical Magazine and Journal of Science*, 10: 1–18.

Venn, J., 1881, *Symbolic Logic*, London: Macmillan.

Venn, J., 1971. *Symbolic Logic*, 2nd edition, New York: Burt Franklin.

Verboon, A.R., 2010, *Lines of Thought: Diagrammatic representation and the Scientific Texts of the Arts Faculty, 1200–1500*, PhD thesis, Leiden University.

Wang, D, Lee, J., and Zeevat, H., 1995, "Reasoning with Diagrammatic Representations," in: Glasgow, Narayanan, and Chandrasekaran.

Wesoły, M., 1996, "Aristotle's Lost Diagrams of the Analytical Figures," *Eos*, 84: 53–64.

Wesoły, M., 2012, "ΑΝΑΛΥΣΙΣ ΠΕΡΙ ΤΑ ΣΧΗΜΑΤΑ: Restoring Aristotle's Lost Diagrams of the Syllogistic Figures," *Peitho*, 3: 83–114.

Westerståhl, D., 1989, "Aristotelian Syllogisms and Generalized Quantifiers," *Studia Logica*, 48: 577–585.

Wetherick, N.E., 1989, "Psychology and Syllogistic Reasoning," *Philosophical Psychology*, 2: 111–124.

Wieb, P., 1990–91, "Existential Assumptions for Aristotelian Logic," *Journal of Philosophical Research*, 16: 321–328.

Williamson, C., 1971, "Traditional Logic as a Logic of Distribution Values," *Logique et Analyse*, 14: 729–746.
Williamson, C., 1972, "Squares of Opposition: Comparisons between Syllogistic and Propositional Logic," *Notre Dame Journal of Formal Logic*, 13: 497–500.
Wilson, N., 1987, "The Distribution of Terms: A Defense of the Traditional Doctrine," *Notre Dame Journal of Formal Logic*, 28: 439–454.
Wittgenstein, L., 1921, *Tractatus Logico-Philosophicus*, B. Pears and B. McGuinness (trans), London: Routledge & Kegan Paul (1961).
Woods, J., 2014, *Errors of Reason: Naturalizing the Logic of Inference*, London: College Publications.
Zemach, E.M., 1981, "Names and Particulars," *Philosophia*, 10: 217–223.
Zemach, E.M., 1985, "On Negative Names," *Philosophia*, 15: 139–138.
Zeman, J.J., 1964, *The Graphical Logic of C. S. Peirce*, Chicago: University of Chicago.

Index

abstract 62, 71, 76, 101, 145, 147, 154, 189
accessibility 142–144, 147
accuracy 55
actual 26, 54, 77, 126, 153, 158, 192
addition/subtraction 48, 72f., 75, 118, 130, 140, 142, 154, 169, 173, 180, 187
adicity 114
affirm/affirmative/affirmation 13, 14,64, 65, 74, 78, 79, 80, 81, 82, 83, 88, 89, 92, 104, 118, 125
algebra 3, 9–11, 66, 72, 124, 164, 192
algebraic 35, 37f., 41, 61, 90, 137, 163, 165, 170, 174, 181, 185
Alpha 42f., 47, 124
Alvarez, E. 87, 89
anaphoric/cataphoric 66, 73, 119
Aristotle 5f., 8, 12–22, 24–26, 30, 56, 58, 66–72, 76, 81f., 86f., 89–91, 98, 101, 106, 124, 129, 148, 157f., 160, 163, 194
Aristotle's First Square 15
Aristotle's lost diagrams 25f.
Aristotle's Second Square 15
Aristotlian/Aristotle's Proviso 87
as a special branch of term logic 129
asymmetry thesis 99–101, 103
axiomatic 21
Azzouni, J. 188

Baron, M. 28
Barwise, J. 54f., 147, 152, 157
Bauer, M. 153, 158f.
bearer of 139
Begriffsschrift/concept-script 34, 48, 49, 132
Bellucci, F. 29, 32, 39
Ben-Yami, H. 14
Beta 42f., 47, 149, 151f.
Beta Graphs 43–46, 124, 148–150
binary 16, 72f., 75, 103, 113f., 186
biology 155, 160f.
Bird, O. 98
Blackwell, A.F. 153
Blake, A. 143f., 192
Bocheński, I.M. 70, 106

Boole, G. 13, 35, 37f., 41, 44, 48f., 66, 74, 92, 124, 192
Boulet, G. 154
Brady, G. 44
Braine, M.D.S. 159
Burch, R.W. 44
Burton, J. 143f.
Byrn, R.J. 158

cancellation 94
Carey, S. 153
Carroll, L. (Charles Dodgson) 13, 38f., 188
Castañeda, H.-N. 70
Castro-Manzano, J.M. 7, 14
Champagne. M. 83
Chapman, P. 192
charge 72, 74, 77f., 93, 99, 117, 154
Cheng, P. C.-H. 143, 153
Chomsky, N. 155
Clark, D.S. 101
clutter 142–147, 150, 152, 192
cognitive 5f., 25, 53, 56, 142f., 147, 150, 153, 156–158, 160, 162, 189
cognitive hypothesis 156
cognitively veridical 156f.
common sense 163, 169f., 172, 174, 177, 181, 185
complete 12, 16f., 21, 26, 38, 42, 55, 68, 78, 91, 96, 165, 190
completeness 58
computational cost 140
computational efficiency 140
connotation 29
constituent 20, 107, 123f., 126–128
constitutive characteristic 127
containment 12, 17, 29, 125
Context Principle 124
contradict/contradictory/contradiction 13, 14, 15, 20, 35, 36, 63, 64, 66, 74, 81, 82, 83, 84, 97, 123, 127, 176, 185
contrapiction 83, 86, 97f., 123, 134f., 138
contrary/contrariety 13, 14, 15, 36, 37, 81, 82, 86, 97, 100, 126, 186

conventionality 55
Cooper, W.S. 155f.
copula 15–19, 22f., 62, 65, 68f., 71–74, 106f.
copulation 68, 107
Corcoran, J. 5, 14, 21, 92, 123, 126, 158
Correia, M. 87, 89
Cox, R. 153
cut 42f., 45f.

Danka, I. 54f., 139
definite description 66, 73
delineated 76, 81, 88, 90, 129, 130
demonstrate/demonstration 5, 20
De Morgan, A. 1, 44, 58, 62, 64f., 70–72, 74, 111, 137
Demy, L. 14
denial 14, 64f., 80, 167
denotation 78f., 89, 100, 103f., 116–118, 126–128
denote object 126
De Rijke, M. 140, 145
Descartes, R. 76, 160, 162
diagrammatic 5f., 12f., 22, 24, 31f., 34f., 38, 44, 47f., 54–56, 58, 62, 64–66, 75, 77, 84, 88, 90f., 103, 112, 136–141, 143, 145–148, 150, 163, 176–178, 180–182, 188–192
diagrammatology 5–7, 54–56, 58, 139
dictum de omni et nullo 92, 93, 94, 137
Dipert, R. 124
direct 1, 5, 47, 75, 113–115, 176, 185, 191
directed graph 58–61
disanalogy 127f.
distribution 28f., 65, 89, 168f., 186
distribution value 29, 31f., 65, 89f., 92, 99
domain 48, 77, 82, 100, 117, 119, 125–132, 136f.
domain of domains 131

ED 7, 66, 77, 79, 86, 88, 90f., 93, 95f., 98, 105f., 111, 115, 129f., 132, 137, 145–152, 191f.
efficacy 140, 142f., 147, 192
ekthesis 25
Elio, R. 153, 158

empty/nonempty 33, 37f., 40, 58, 61, 78, 86–88, 126, 146
Englebretsen Diagram 66
Englebretsen, G. 16, 25, 56, 66–71, 85, 89, 92, 98–100, 103, 105f., 127, 154, 163–165, 172, 176–178, 180, 182, 185f.
enthymeme 65, 87, 111, 147, 153
equivalence (logical) 47, 128, 132
Etchemendy, J. 54f., 147, 152
Euclid 12, 76
Euclidean 54, 117, 146
Euler, L. 1, 4, 6, 25–27, 29f., 32–39, 55, 58, 66, 77, 79, 142, 144–147, 188, 191
evolution 155, 161f.
exclusion 17, 29, 31, 77, 79, 144
existence 33, 38, 87f., 96, 126, 129, 173
existential commitment 86
Existential Graph 40f., 46, 54, 56, 144
existential import 63, 88, 96, 129, 164, 173
explicit 2, 42, 51, 64f., 72, 87f., 91, 96–98, 104f., 128–130, 140, 154, 159f., 190
explicit denotation 103–105
expressive 6, 14, 40, 45f., 56, 74, 142, 145–147, 172
expressiveness 142, 145, 192

fact 19, 26f., 30, 32, 37, 44, 48, 54, 72–74, 76f., 87–89, 104, 106, 117f., 124f., 127, 129, 137, 154, 156, 159, 164, 169, 171, 180, 185, 189f.
figure (syllogistic) 1f., 4–6, 8–13, 15, 17–19, 21–31, 33–47, 49–55, 58–61, 63f., 77–81, 83–86, 88, 90–99, 101–103, 105–123, 129–138, 143, 145–149, 151f., 154, 164f., 167f., 170f., 173, 176–185, 190f.
first-order 6, 46, 48, 66, 164, 185
Fish, A. 69, 142
focus 13, 56, 61, 76, 153, 161, 164, 166, 190
formal 2f., 5–7, 13–16, 20, 25f., 38, 41, 45, 48, 50, 53–55, 61, 65f., 71f., 74f., 88, 90f., 103–106, 123–125, 127, 155, 174, 188f.
Francavilla, M. 154
Franklin, J. 5
Frederick, D. 14, 103

free ride 141–143, 191
Frege, G. 6, 34, 42, 47–55, 67f., 99, 124, 126, 132, 157f.
Friedman, W.H. 89
functor 72, 132

Gaines, B. 14
Gardner, M. 38, 85, 163
Gödel, K. 54
Geach, P. 13, 16, 89, 186
Giaquinto, M. 153
Giardino, V. 189f., 192
Golden Ratio 9f.
Gottfried, B. 143–145
graphic 1, 5, 13, 35, 41f., 44–47, 51f., 54, 59, 61–63, 66, 77, 79, 88, 106, 129, 132, 139, 147, 149, 160, 188, 191f.
Green, J. 13, 81f., 92, 100
Gurr, C. 142, 191

Hamami, Y. 160
Hammer, E. 34, 38, 41
Hanson, P. 153, 158
Hellman, G. 76
Helly's Theorem 146, 150, 191
heterogeneous 42, 55f., 61f., 68, 75, 147
hidden premise 96, 112
Hobbes, T. 72
Hodges, W. 89f., 126
Hoffmann, M.G.H. 153
homological 55f.
homomorphic 143
Hornbeck, C. 14
Howse, J. 142f.
Hume, D. 126, 152

icon 42, 46f., 78, 143, 157
iconicity 142f., 147, 150
identity/ identical/'is' of identity 37, 43, 45, 46
immediate 55, 87, 92, 114, 141, 150, 186
imperfect 19f., 24f., 63, 92, 94
impossible 12, 20, 25, 82f., 100f.
inclusion 31, 77, 79, 81, 110, 123, 144, 192
incompatibility group 100
incompatible 100, 127f., 138
incomplete 37, 40, 67, 92

indefinite noun 14
Index 171f., 180
indirect 20, 64, 113–115
individual 27, 29, 31f., 37, 46, 71, 73, 77f., 80f., 99–102, 104, 113, 115, 130, 144–146, 149f., 158f., 164
inference 3, 5–7, 12, 25, 27f., 35, 47, 51–55, 62f., 66–68, 70, 72, 75, 86f., 91f., 103, 111f., 114, 120–125, 130, 132, 137, 139–145, 147, 150, 153, 157–159, 163, 165, 170, 173, 181, 186, 190f.
inferential 14, 53, 87, 142, 147, 163, 170, 173, 185
innate 7, 155f., 158, 160
interpretation 13, 25, 29f., 38, 41, 89, 124
intersection 5, 40, 77, 79, 81, 88, 130, 144, 146, 180, 182, 192
in the domain vs of the domain 127, 128
intuitive 188
Izard, V. 153

Jamnik, M. 143
John, C. 5, 14, 35, 55, 103–105, 123, 142
Johnson-Laird, P. N. 153, 158f., 185

Kanterian, E. 52
Kant, I. 4, 26, 30, 74, 124, 126
Katz, B.D. 89
Khemlani, S. 185
Kneale, M. 87, 92
Kneale, W. 87, 92
Kuhn, S.T. 164

Lambert, J.H. 6, 26f., 30–33, 56, 58, 61, 77, 191f.
Larkin, J. 139–142, 159
Laronge, J.A. 14
laws of 50, 63, 155–158, 160
Lee, J. 153, 191
Lee, S.A. 153, 191
Legg, C. 160, 188f.
Lehan-Streisel, V. 158
Leibniz, G.W. 5f., 8, 13, 26–33, 49, 53, 56, 58, 61, 66, 70–72, 74, 77, 92, 98, 102, 106, 124f., 127, 137, 148, 191f.
Leibniz-Sommers logical program 127

Lemon, A 53, 140, 143, 145 f., 148–150, 153, 159
Lemon and Pratt Counterexample/challenge 145, 146, 148, 149, 150
Lemon, O. 53, 140, 143, 145 f., 148–150, 153, 159
lexicon/logical vocabulary 16, 22, 42, 43, 47, 68, 178
Lindsay, R.K. 159
Linnebo, Ø. 76
Lockwood, M. 103
logic 3–8, 13–16, 20–22, 25–27, 30, 32, 34 f., 37 f., 41 f., 44, 46–51, 53–56, 58, 61 f., 65–72, 74–77, 86, 88, 90, 92, 96, 98 f., 101, 105 f., 111–113, 117, 123–125, 127–133, 137, 139, 141–143, 145, 147, 150, 152, 155–159, 163 f., 185 f., 189, 191–193
logical 1, 4–6, 8, 12–15, 17, 19–21, 23, 25–27, 32, 34 f., 39, 41 f., 44, 47–51, 54–56, 58, 61 f., 64–75, 77 f., 82, 85 f., 88, 98 f., 103 f., 107, 113, 117 f., 121, 123–125, 129 f., 132, 139, 143, 145, 147, 150, 153, 155 f., 158, 160, 164 f., 185 f., 188, 192
logical syntax 15 f., 19, 67–69, 71, 132, 156
lost premise 91
Lowe, R.K. 55, 143, 153
Łukasiewicz, J. 87, 98

Macbeth, D. 48, 190
Makinson, D. 89
Marquand, A. 39
Martinich, A.P. 89
Martin, J.N. 7, 14, 89
Masthoff, J. 34, 143
mathematical 3, 5, 8, 25, 48 f., 51, 53, 55, 67, 139, 190
mediate 68, 87, 92, 137, 150, 186
medieval 23, 25, 29, 69–71, 87, 117, 137
mental model 7, 159
mental sentence 159
Merrill, D.D. 44
Miallef, L. 192
Mineshima, K. 14, 34, 79, 143–145
minimizing 85, 142
Mirzapour, M. 14

modern 3, 5 f., 25, 42, 46, 76, 86 f., 99, 105 f., 113, 117, 139, 150, 157
modern predicte logic 6, 42, 87, 113
Moktefi, A. 5, 14, 26–30, 32, 39 f., 47, 50, 54, 56, 103, 143, 163, 189, 191
Morrissey, C. 14
Moss, L. 14, 164
Mostowski, A. 185
MPL 42, 71, 104, 114, 117, 119, 124, 130, 155
Mumma, J. 160
Murphree, W. 14, 185

name 2, 46, 66, 68, 73, 98 f., 104, 113
Natali, C. 13
natural 2–4, 14, 21, 45, 50, 55, 67, 69–75, 98, 114, 119, 139, 141, 143, 150, 152 f., 155 f., 158, 160–162, 164, 192
negation 14, 39, 42 f., 47 f., 64–66, 74, 80–83, 90, 100, 127, 132, 146, 153, 186
negative 13 f., 28, 39, 58, 61, 65, 73–75, 78–83, 86 f., 89 f., 93, 99–101, 113, 121–123, 127, 137, 145, 154, 163, 166–169, 171, 186, 191
neo-Aristotelian 161
Noah, A. 71, 103, 164 f.
nonlogical 82
non-symbolic 55
Nùñez, R. 154

Oberlander, J. 153, 160
object 55, 62, 70 f., 74, 76, 78 f., 81, 99, 104, 113–118, 126 f., 130, 143, 153 f., 157, 160, 176, 178, 182, 189
O'Brien, D.P. 159
Oderberg, D. 14
Okada, M. 79, 143
operational constraint 141 f., 150
overdetermined alternative 141 f., 145 f., 150, 191
overlap 77

Pacheco-Montes, J.R. 7, 14
Pagnan, R. 58, 61–66, 87
paradox 54 f.
Parsons, T. 69 f., 87, 89
Patzig, G. 98
Peano, G. 48

218 — Index

pegasizing 46
Peirce, C.S. 6, 13, 34, 38, 40–42, 44, 46 f., 53–56, 58, 71, 74, 76–78, 124, 139, 143 f., 148, 152–154, 188, 191
Pelletier, F.J. 153, 158
perceptual 55, 140 f., 153, 157, 188 f.
perfect 19–21, 24 f., 58, 60, 63, 91–93, 145, 191
perspicacity 25, 39, 55
perspicuity 56
perspicuous 35, 37, 39, 47, 50, 71, 96, 106, 188 f., 192
Peterson, P.L. 14, 163, 166, 185
phrasal conjunction 103 f.
phrasal disjunction 103 f.
Pietarinen, A.-V. 14, 28 f., 32, 39 f.
Plato 4 f., 8, 12, 16, 67, 73
plus-minus 163
plus-minus algebra 163–165, 170–172, 175 f., 185
post-Fregean 6, 16, 70, 124
power 5 f., 14, 20, 45 f., 56, 74, 115, 141 f., 145, 147, 161, 172
practice based 7, 189, 190
Pratt, I. 145 f., 148–150
predicate 3, 6, 13 f., 28, 38, 41–44, 46, 48–51, 65–68, 71, 73, 80, 86 f., 92, 99–101, 107, 113 f., 163, 168, 175, 186 f.
predicate functor 71
Predicate Functor Algebra (PFA) 71
predicate term 14, 16–18, 21 f., 24, 27, 29, 31, 73, 80–82, 89, 92, 99, 101–103, 107, 111, 121 f., 129, 182
presence/absence 126, 127, 132, 142, 144, 153
primary 13, 22, 37, 51, 56, 70, 72, 101, 124, 143, 146
primary logic 49, 124, 132
Principle of Compound Term Analysis 147
Principle of Noncontradiction 81–83
Principle of Relational Analysis 112, 147
Principle of Relational Extension 112
Principle of Relational Reduction 114
Priority (Context) Principle 49, 124
privative 82
process 26, 51, 75, 85, 91, 139–142, 155 f., 159 f., 188 f.

pronoun/pronominalization 113, 115, 116, 118, 119, 120
proper 5, 66, 73, 77, 79, 116, 137, 157, 189
proposition 8, 13, 16, 20, 24, 27, 38, 42, 47, 49–51, 58 f., 62, 70–72, 74, 85, 90, 92, 100, 102, 123–128, 132 f., 137, 139, 163–169, 171–174, 180–182
propositional 38, 41 f., 49, 51, 74, 77, 92, 123 f., 127–130, 132 f., 165
propositional unity 67
proto-statement 17–19, 22
psuché 160
psyche 160–162
psychologically veridical 156
psychologism 157 f.
Putnam, H. 76

Quine, W.V. 46, 71, 75, 79, 126, 164

Ramsey, F. 99 f.
ratiocination 143, 147, 156, 159, 188, 192
rational 8, 17 f., 74, 139, 155 f., 160–162, 171, 189 f., 192
Rauf, J.V. 66
Rearden, M. 89
reasoning 7, 13, 21, 26, 33 f., 48–50, 54–56, 72, 76, 78, 139, 141 f., 144, 146, 152 f., 155–163, 165 f., 169 f., 172–174, 177, 180–182, 185, 189–192
reflexive 103, 115–117, 125
representation 1, 5, 14 f., 28, 32 f., 35, 37, 39, 42, 45, 47, 49, 51–53, 55 f., 58, 61 f., 66, 76 f., 79 f., 84 f., 90, 106, 109, 132, 139–146, 150, 152, 159 f., 165 f., 169, 171–174, 177, 180–182, 190 f.
rigor 55
Rips, L. 156, 159
Roberts, D.D. 41, 149
Roeper, P. 76
Rogers, P. 143, 192
Ross, W.D. 8, 12 f., 16, 98
Russell, B. 5, 13 f., 47 f., 54, 76, 104, 114, 123, 126, 188

Sato, Y. 14, 34, 79, 143–145
saturated/unsturated 67 f.
Savio, M. 34

Sayward, C.   67, 186
Scaife, M.   153
Schang, F.   14
Schlimm, D.   132
Schröder, E.   48
Selçuk, T.   14
semantic relation   125 f., 142
sentential term   74, 124, 127–129, 131, 133, 137, 150
Shapiro, S.   76
sheet of assertion   42
Sheffer, H.M   42
Shimojima, A.   140–143, 145
Shin, S.-J.   14, 26 f., 30, 34–37, 40–42, 44, 47, 50, 54–56, 105, 163
signify   126, 127
Simon, H.A.   139–142, 159
simplicity   39, 57, 145, 147, 150, 152
singular term   6, 13, 31, 61, 66, 69, 71, 73 f., 78, 92, 98–104, 113, 127, 129, 137, 148 f., 165, 191
Slater, B.H.   14, 46
Smiley, T.J.   2, 21
Smit, H.   161 f.
Smith, R.   12
Smyth, M.B.   58–62
Sommers, F.   5 f., 13 f., 16, 66 f., 70–72, 74, 89 f., 92, 99, 103, 106, 113, 117 f., 123–129, 155–157, 163 f., 172, 185, 191
sorites   96
sound/soundness   21, 38, 42, 55
specific vs non-specific reference   113, 118, 130
Spelke, E.   153
spicular notation   64
split/unsplit copula   69, 72, 74, 106, 107
Square of Opposition   13–15, 26, 85 f., 88, 167 f.
Stapleton, G.   34, 143 f., 192
Stenning, K.   145 f., 153, 159 f., 191
Stjernfelt, F.   139
Stoic   124
Strawson, P.F.   100 f., 103
strengthened   13 f., 63, 65, 70 f., 150
subject   13 f., 26, 29, 31, 34, 54, 66 f., 70 f., 73, 82, 92, 98–101, 107, 113, 115–118, 125 f., 129 f., 143 f., 153, 159, 163, 168, 175, 182, 186–188, 190, 192
subject term   16–18, 21 f., 24, 27–29, 31, 73, 87, 89, 92, 98 f., 101–104, 106, 111, 115, 128 f., 148, 163, 183
subscribed numerals   106
substitution   51, 63, 92, 137
suitedness   142
supposition   89, 126
SYLL   62–66, 87, 163
SYLL$^+$ 166, 168, 170–172, 174–176
SYLL$_{valid}$ 175, 176
Syllogism   1, 8, 12–14, 16–22, 24–28, 31–33, 36 f., 39 f., 58, 60, 62 f., 65, 70 f., 75, 87 f., 90–96, 98, 102, 106, 123, 130, 144, 147, 150, 152, 158, 164 f., 174–177, 180–182, 185, 191
syllogistic   6, 12, 14–29, 33 f., 41, 44, 49, 58, 62–70, 86 f., 90–93, 96, 98, 106, 123, 137, 144, 147, 150, 158, 163–166, 169–177, 181, 184 f.
symbolic   2 f., 13, 20, 38, 41 f., 47–49, 51, 53–56, 66, 69, 72, 74 f., 123, 127, 132, 139, 144, 147, 150, 159, 188
syncategoremata   69. 72
syntax (logical form)   47, 74, 155 f., 164, 170, 176 f., 179
Szabó, Á.   8
Szabolcsi, L.   14

Takemura, R.   34, 79, 143 f.
Taylor, J.   142
team name   104 f.
Tennant, N.   5
term   5 f., 8, 12–22, 24–32, 35, 37–40, 46 f., 49, 55 f., 58 f., 61–84, 86–90, 92–94, 96–107, 110, 113–118, 121, 123–130, 132 f., 137 f., 142, 145, 147, 150, 153–156, 158, 161, 163–165, 168 f., 171, 173, 175 f., 180, 182 f., 185–187, 191
Term Functor Logic   5 f., 13, 66 f., 69, 106, 156, 164, 191
ternary   16, 113
TFL   5, 7, 67, 69, 71 f., 74 f., 80, 83, 87, 90, 92, 96, 98 f., 101–106, 111, 113 f., 117 f., 121, 125, 127, 129 f., 132, 137, 148, 150, 155 f., 164, 170–172, 176, 182, 186, 191 f.

TFL⁺ 170, 173–178, 180, 182, 184f.
TFL^⊕ 176–185
TFL_valid 175, 176, 182, 183
theory of 15f., 20, 25, 42, 51, 68f., 89, 101, 103, 132, 155, 159, 174, 181
Thompson, B. 163, 166–168, 170–172, 185
Toader, I.D. 48
totality 127
Touloumis, A. 143f.
traditional 14, 19, 44, 49, 64, 66, 69f., 80, 85–88, 90, 92, 98f., 105f., 117, 137, 150, 156, 174, 181
Tredennick, H. 13, 19
truth claim 128
truth-value 38, 68
Tweaked 170, 172–174, 185

Ueda, K. 143
universe of discourse 3, 37–39, 42, 78, 87, 100, 117, 126

valid 19–21, 24, 52, 58, 63, 88, 91, 96f., 103, 111, 116, 132, 150, 153, 156, 164f., 169f., 172–177, 180–185

vector 6, 26, 77, 105–107, 109
Venn, J. 6, 13, 25, 34f., 37–41, 47, 53, 56, 58, 66, 77, 79, 96, 144f., 147, 188, 191f.
verbaliser vs visualiser 159
visual 1, 5, 15, 28, 35, 37, 41, 52–56, 66, 75, 129, 140f., 144f., 152–155, 189f.
visual inference/visually infer 159

Wajima, Y. 143
weakened 19, 63, 96
well-matchedness 142f., 145, 147
Wesoły, M. 14, 21–26, 58
Westerståhl, D. 185
wild quantity thesis 102, 128
Williamson, C. 89
Wilson, N. 89
Wittgenstein, L. 124, 157
Wood, S. 92
world 77, 124, 126–131, 159
world line 123

Zemach, E.M. 101
Zeman, J.J. 41

www.ingramcontent.com/pod-product-compliance
Lightning Source LLC
Chambersburg PA
CBHW031812220426
43662CB00007B/605